Plant Responses to Environmental Stress

Plant Responses to Environmental Stress

EDITED BY

M.F. Smallwood, C.M. Calvert and D.J. Bowles
The Plant Laboratory, The University of York, PO Box 373, York YO1 5YW, UK

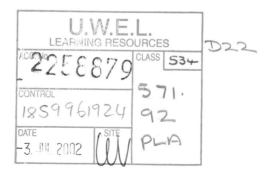

βIOS
SCIENTIFIC
PUBLISHERS

© BIOS Scientific Publishers Limited, 1999

First published in 1999

A CIP catalogue record for this book is available from the British Library.

ISBN 1 85996 192 4

BIOS Scientific Publishers Ltd
9 Newtec Place, Magdalen Road, Oxford OX4 1RE, UK
Tel. +44 (0) 1865 726286. Fax. +44 (0) 1865 246823
World Wide Web home page: http://www.bios.co.uk/

Production Editor: Jonathan Gunning.
Typeset by Saxon Graphics Ltd, Derby, UK.
Printed by Biddles Ltd, Guildford, UK.

Contents

SECTION 3: Reactive Oxygen Species in Abiotic Stress Responses

SECTION 4: Temperature Stress

Contributors

Aalen, R.B. Division of General Genetics, University of Oslo, P.O. Box 1031 Blindern, N-0315 Oslo, Norway

Abe, H. Biological Resources Division, Japan International Research Center for Agricultural Sciences (JIRCAS), 2–1 Ohwashi, Tsukuba, Abaraki 305–0851 Japan

Alonso, J.M. Plant Science Institute, Department of Biology, University of Pennsylvania, Philadelphia, PA 19104-6018, USA

Altabella, T. Unitat de Fisiologia Vegetal, Facultat de Farmacia, Universitat de Barcelona, Avda Diagonal 643, 08028 Barcelona, Spain

Anegg, S. GSF – Institute of Biochemical Plant Pathology, D-85764 Neuherberg, Germany

Bartels, D. Institut für Botanik, Universität Bonn, Kirschallee 1, D-53115 Bonn, Germany

Bar-Zvi, D. Department of Life Sciences, Ben-Gurion University, Beer-Sheva 84105, Israel

Beeching, J.R. Department of Biology and Biochemistry, University of Bath, Bath BA2 7AY, UK

Ben-Hayyim, G. Department of Fruit Tree Breeding and Molecular Genetics, Agricultural Research Organization, The Volcano Center, P.O. Box 6, 50250 Bet Dagan, Israel

Blomstedt, C.K. Department of Biological Sciences, Monash University, Clayton 3168, Victoria, Australia

Bortolotti, C. Unitat de Fisiologia Vegetal, Facultat de Farmacia, Universitat de Barcelona, Avda Diagonal 643, 08028 Barcelona, Spain

Bowley, S.R. Plant Biotechnology Division, Department of Plant Agriculture, Crop Science Building, University of Guelph, Guelph, Ontario N1G 2W1, Canada

Brown, A.P.C. Department of Plant Sciences, University of Cambridge, Downing Street, Cambridge CB2 3EA, UK

Buschmann, H. Department of Biology and Biochemistry, University of Bath, Bath BA2 7AY, UK

Clarke, A. Department of Biological and Biomedical Sciences, University of the West of England, Coldharbour Lane, Bristol BS16 1QY, UK

Cordeiro, A. Unitat de Fisiologia Vegetal, Facultat de Farmacia, Universitat de Barcelona, Avda Diagonal 643, 08028 Barcelona, Spain

Creissen, G. John Innes Centre, Norwich Research Park, Colney, Norwich NR4 7UH, UK

Cubells, X. Instituto de Biología Molecular y Celular de Plantas, CSIC, Universidad Politécnica, Camino de Vera s/n, E-46022 Valencia, Spain

Desikan, R. Department of Biological and Biomedical Sciences, University of the West of England, Coldharbour Lane, Bristol BS16 1QY, UK

Dunn, M.A. Department of Biochemistry and Genetics, The Medical School, University of Newcastle, Newcastle upon Tyne NE2 4HH, UK

Ernst, D. GSF – Institute of Biochemical Plant Pathology, D-85764 Neuherberg, Germany

Eshdat, Y. Department of Fruit Tree Breeding and Molecular Genetics, Agricultural Research Organization, The Volcano Center, P.O. Box 6, 50250 Bet Dagan, Israel

Fourcroy, P. Biochimie et Physiologie Moléculaire des Plantes, CNRS URA 2133, ENSAM-INRA, 2 place Viala, F-34060 Montpellier cedex 01, France

Frazão, A.S. National Center of Research on Food Technology, Avenida das Américas 29.501 CEP 23.020-470 Rio de Janeiro-Brasil

Gaff, D.F. Department of Biological Sciences, Monash University, Clayton 3168, Victoria, Australia

Ghasempour, H.R. Biology Department, Razi University, Shahid Beheshti Boulevard, Kermanshah 67147, Iran

Gianello, R.D. Department of Biochemistry and Molecular Biology, Monash University, Wellington Road, Clayton 3168, Victoria, Australia

Gilad, A. Department of Life Sciences, Ben-Gurion University, Beer-Sheva 84105, Israel

Gillet, B. CEA/Cadarache, DSV, DEVM, Laboratoire d'Ecophysiologie de la Photosynthèse, F-13108 Saint-Paul-lez-Durance Cedex, France

Gil-Mascarell, R. Instituto de Biología Molecular y Celular de Plantas, Universidad Politécnica de Valencia – Consejo Superior de Investigaciones Científicas, Camino de Vera, E-46022 Valencia, Spain

Gómez-Vásquez, R. Department of Biology and Biochemistry, University of Bath, Bath BA2 7AY, UK

Gossen, B. Agriculture and Agri-Food Canada, Saskatoon Research Centre, 107 Science Place, Saskatoon, Saskatchewan S7N 0X2, Canada

Granell, A. Instituto de Biología Molecular y Celular de Plantas, CSIC, Universidad Politécnica, Camino de Vera s/n, E-46022 Valencia, Spain

Grimmig, B. GSF – Institute of Biochemical Plant Pathology, D-85764 Neuherberg, Germany

Hamill, J.D. Department of Biological Sciences, Monash University, Clayton 3168, Victoria, Australia

Han, Y. Department of Biology and Biochemistry, University of Bath, Bath BA2 7AY, UK

Hancock, J. Department of Biological and Biomedical Sciences, University of the West of England, Coldharbour Lane, Bristol BS16 1QY, UK

Haslekås, C. Division of General Genetics, University of Oslo, P.O. Box 1031 Blindern, N-0315 Oslo, Norway

Heidenreich, B. GSF – Institute of Biochemical Plant Pathology, D-85764 Neuherberg, Germany

Hincha, D.K. Institut für Pflanzenphysiologie und Mikrobiologie, Freie Universität, Königin Luise Str. 12–16, D-14195 Berlin, Germany

Holland, D. Department of Fruit Tree Breeding and Molecular Genetics, Agricultural Research Organization, The Volcano Center, P.O. Box 6, 50250 Bet Dagan, Israel

Holt, C.B. Unilever Research Laboratory, Colworth House, Sharnbrook, Bedfordshire MK44 1LQ, UK.

Hughes, M.A. Department of Biochemistry and Genetics, The Medical School, University of Newcastle, Newcastle upon Tyne NE2 4HH, UK

Ichimura, K. Laboratory of Plant Molecular Biology, Tsukuba Life Science Center, The Institute of Physical and Chemical Research (RIKEN), 3–1–1 Koyadai, Tsukuba, Ibaraki 305–0074 Japan

Iglesias, C. CIAT, A.A. 6713, Cali, Colombia

Jenkins, G.I. Plant Molecular Science Group, Division of Biochemistry and Molecular Biology, Institute of Biomedical and Life Sciences, Bower Building, University of Glasgow, Glasgow G12 8QQ, UK

Ito, T. Laboratory of Plant Molecular Biology, Tsukuba Life Science Center, The Institute of Physical and Chemical Research (RIKEN), 3–1–1 Koyadai, Tsukuba, Ibaraki 305–0074 Japan

Jones, K.S. Plant Biotechnology Division, Department of Plant Agriculture, Crop Science Building, University of Guelph, Guelph, Ontario N1G 2W1, Canada

Karpinska, B. Department of Forest Genetics and Plant Physiology, Faculty of Forestry, Swedish University of Agricultural Sciences, SE-901 83 Umeå, Sweden

Karpinski, S. Department of Forest Genetics and Plant Physiology, Faculty of Forestry, Swedish University of Agricultural Sciences, SE-901 83 Umeå, Sweden

Kasuga, M. Biological Resources Division, Japan International Research Center for Agricultural Sciences (JIRCAS), 2–1 Ohwashi, Tsukuba, Ibaraki 305–0851 Japan

Knight, H. Department of Plant Sciences, University of Oxford, South Parks Road, Oxford OX1 3RB, UK

Knight, M.R. Department of Plant Sciences, University of Oxford, South Parks Road, Oxford OX1 3RB, UK

Kuang, J. Monell Chemical Senses Center, 3500 Market Street, Philadelphia, PA 19104, USA

Lafuente, M.T. Instituto de Agroquímica y Tecnología de Alimentos, CSIC, Apdo Postal 73, E-46100 Burjassot, Spain

Langebartels, C. GSF – Institute of Biochemical Plant Pathology, D-85764 Neuherberg, Germany

Li, H. Department of Biology and Biochemistry, University of Bath, Bath BA2 7AY, UK

Lillford, P.J. Unilever Research Laboratory, Colworth House, Sharnbrook, Bedfordshire MK44 1LQ, UK

Lima, F.H. National Center of Research on Food Technology, Avenida das Américas 29.501 CEP 23.020-470 Rio de Janeiro-Brasil

Liu, Q. Biological Resources Division, Japan International Research Center for Agricultural Sciences (JIRCAS), 2–1 Ohwashi, Tsukuba, Ibaraki 305–0851 Japan

López-Coronado, J.M. Instituto de Biología Molecular y Celular de Plantas, Universidad Politécnica de Valencia – Consejo Superior de Investigaciones Científicas, Camino de Vera, E-46022 Valencia, Spain

Mackerness, S. A.-H. Horticulture Research International, Wellesbourne, Warwickshire CV35 9EF, UK

Masgrau, C. Unitat de Fisiologia Vegetal, Facultat de Farmacia, Universitat de Barcelona, Avda Diagonal 643, 08028 Barcelona, Spain

McKersie, B.D. Plant Biotechnology Division, Department of Plant Agriculture, Crop Science Building, University of Guelph, Guelph, Ontario N1G 2W1, Canada

Medina, J. Departamento de Mejora Genética y Biotecnología, INIA, Carretera de la Coruña, Km 7, E-28040 Madrid, Spain

Miyata, S. Biological Resources Division, Japan International Research Center for Agricultural Sciences (JIRCAS), 2–1 Ohwashi, Tsukuba, Ibaraki 305–0851 Japan

Mizoguchi, T. Laboratory of Plant Molecular Biology, Tsukuba Life Science Center, The Institute of Physical and Chemical Research (RIKEN), 3–1–1 Koyadai, Tsukuba, Ibaraki 305–0074 Japan

Moeder, W. GSF – Institute of Biochemical Plant Pathology, D-85764 Neuherberg, Germany

Mullineaux, P. John Innes Centre, Norwich Research Park, Colney, Norwich NR4 7UH, UK

Murguía, J.R. Instituto de Biología Molecular y Celular de Plantas, Universidad Politécnica de Valencia – Consejo Superior de Investigaciones Científicas, Camino de Vera, E-46022 Valencia, Spain

Nakashima, K. Biological Resources Division, Japan International Research Center for Agricultural Sciences (JIRCAS), 2–1 Ohwashi, Tsukuba, Ibaraki 305–0851 Japan

Neale, A.D. Department of Biological Sciences, Monash University, Clayton 3168, Victoria, Australia

Neill, S. Department of Biological and Biomedical Sciences, University of the West of England, Coldharbour Lane, Bristol BS16 1QY, UK

Neukamm, B. Institut für Pflanzenphysiologie und Mikrobiologie, Freie Universität, Königin Luise Str. 12–16, D-14195 Berlin, Germany

Panicot, M. Unitat de Fisiologia Vegetal, Facultat de Farmacia, Universitat de Barcelona, Avda Diagonal 643, 08028 Barcelona, Spain

Peltier, G. CEA/Cadarache, DSV, DEVM, Laboratoire d'Ecophysiologie de la Photosynthèse, F-13108 Saint-Paul-lez-Durance Cedex, France

Prändl, R. ZMBP – Zentrum für Molekular Biologie der Pflanzen, Allgemeine Genetik, Universität Tübingen, Auf der Morgenstelle 28, D-72076 Tübingen, Germany

Pruvot, G. CEA/Cadarache, DSV, DEVM, Laboratoire d'Ecophysiologie de la Photosynthèse, F-13108 Saint-Paul-lez-Durance Cedex, France

Reilly, K. Department of Biology and Biochemistry, University of Bath, Bath BA2 7AY, UK

Rey, P. CEA/Cadarache, DSV, DEVM, Laboratoire d'Ecophysiologie de la Photosynthèse, F-13108 Saint-Paul-lez-Durance Cedex, France

Reynolds, H. John Innes Centre, Norwich Research Park, Colney, Norwich NR4 7UH, UK

Rodriguez, M.X. Department of Biology and Biochemistry, University of Bath, Bath BA2 7AY, UK

Rodríguez, P.L. Instituto de Biología Molecular y Celular de Plantas, Universidad Politécnica de Valencia – Consejo Superior de Investigaciones Científicas, Camino de Vera, E-46022 Valencia, Spain

Ruiz, O. Unitat de Fisiologia Vegetal, Facultat de Farmacia, Universitat de Barcelona, Avda Diagonal 643, 08028 Barcelona, Spain

Sakuma, Y. Biological Resources Division, Japan International Research Center for Agricultural Sciences (JIRCAS), 2–1 Ohwashi, Tsukuba, Abaraki 305–0851 Japan

Salinas, J. Departamento de Mejora Genética y Biotecnología, INIA, Carretera de la Coruña, Km 7, E-28040 Madrid, Spain

Sanchez-Ballesta, M.T. Instituto de Agroquímica y Tecnología de Alimentos, CSIC, Apdo Postal 73, E-46100 Burjassot, Spain

Sandermann Jr, H. GSF – Institute of Biochemical Plant Pathology, D-85764 Neuherberg, Germany

Schmitt, J.M. Institut für Pflanzenphysiologie und Mikrobiologie, Freie Universität, Königin Luise Str. 12–16, D-14195 Berlin, Germany

Schöffl, F. ZMBP – Zentrum für Molekular Biologie der Pflanzen, Allgemeine Genetik, Universität Tübingen, Auf der Morgenstelle 28, D-72076 Tübingen, Germany

Schubert, R. Faculty of Forest Sciences, Section Forest Genetics, Ludwig Maximilians University of Munich, Am Hochanger 13, D-85354 Freising, Germany

Seki, M. Laboratory of Plant Molecular Biology, Tsukuba Life Science Center, The Institute of Physical and Chemical Research (RIKEN), 3–1–1 Koyadai, Tsukuba, Ibaraki 305–0074 Japan

Serrano, R. Instituto de Biología Molecular y Celular de Plantas, Universidad Politécnica de Valencia – Consejo Superior de Investigaciones Científicas, Camino de Vera, E-46022 Valencia, Spain

Shinozaki, K. Laboratory of Plant Molecular Biology, Tsukuba Life Science Center, The Institute of Physical and Chemical Research (RIKEN), 3–1–1 Koyadai, Tsukuba, Ibaraki 305–0074 Japan

Shinwari, Z.K. Biological Resources Division, Japan International Research Center for Agricultural Sciences (JIRCAS), 2–1 Ohwashi, Tsukuba, Ibaraki 305–0851 Japan

Stacy, R.A.P. GenoVision Cytomics A.S., Gjerdrumsvei 12A, N-0486 Oslo, Norway

Stephan, M.P. National Center of Research on Food Technology, Avenida das Américas 29.501 CEP 23.020-470 Rio de Janeiro-Brasil

Steponkus, P.L. 611 Bradfield Hall, Department of Soil, Crop and Atmospheric Sciences, Cornell University, Ithaca, NY 14853, USA

Thomas, B. Horticulture Research International, Wellesbourne, Warwickshire CV35 9EF, UK

Thomas, G. GSF – Institute of Biochemical Plant Pathology, D-85764 Neuherberg, Germany

Thomashow, M.F. Department of Crop and Soil Sciences, Michigan State University, East Lansing, MI 48824, USA

Tiburcio, A.F. Unitat de Fisiologia Vegetal, Facultat de Farmacia, Universitat de Barcelona, Avda Diagonal 643, 08028 Barcelona, Spain

Tischendorf, G. Institut für Pflanzenphysiologie und Mikrobiologie, Freie Universität, Königin Luise Str. 12–16, D-14195 Berlin, Germany

Uemura, M. Cryobiosystem Research Center, Iwate University, Morioka, 020–8550, Iwate, Japan

Urao, T. Biological Resources Division, Japan International Research Center for Agricultural Sciences (JIRCAS), 2–1 Ohwashi, Tsukuba, Ibaraki 305–0851 Japan

Vural, S. Department of Biochemistry and Genetics, The Medical School, University of Newcastle, Newcastle upon Tyne NE2 4HH, UK

Wingsle, G. Department of Forest Genetics and Plant Physiology, Faculty of Forestry, Swedish University of Agricultural Sciences, SE-901 83 Umeå, Sweden

Yamaguchi-Shinozaki, K. Biological Resources Division, Japan International Research Center for Agricultural Sciences (JIRCAS), 2–1 Ohwashi, Tsukuba, Ibaraki 305–0851 Japan

Zacarias, L. Instituto de Agroquímica y Tecnología de Alimentos, CSIC, Apdo Postal 73, E-46100 Burjassot, Spain

Abbreviations

3-AT	3-amino-1,2,4-triazole
ABA	abscisic acid
ABRE	ABA-responsive element
ACC	1-aminocyclopropane-1-carboxylate
ADC	arginine decarboxylase
AFGP	antifreeze glycoprotein
AFP	antifreeze protein
AOS	active oxygen species
APS	adenosine 5′-phosphosulphate
APX	ascorbate peroxidase
BAPTA	1,2-bis-(2-aminophenoxy)ethane-N,N,N',N'-tetraacetic acid
BSO	buthionine sulphoxime
BTH	benzothiadiazole
CDPK	calcium-dependent protein kinase
CDSP	chloroplastic drought-induced stress protein
CHRC	chromoplast protein C
CHS	chalcone synthase
CPD	cyclobutane pyrimidine dimer
CRT	C-repeat
DCMIB	2,5-dibromo-3-methyl-6-isopropyl-p-benzoquinone
DCMU	3-(3,4-dichlorophenyl)-1,1-dimethylurea
DGDG	digalactosyldiacylglycerol
DOPC	dioleoylphosphatidylcholine
DOPE	dioleoylphosphatidylethanolamine
DRE	drought-responsive element
DSP	desiccation stress protein
EDTA	ethylenediamine tetraacetic acid
EGTA	ethylene glycol bis-(β-aminoethylether) N,N,N,N'-tetraacetic acid
EIL	expansion-induced lysis
EL	excess light
EMSA	electrophoretic mobility shift analysis
EPE	excess photochemical energy
ERE	ethylene-responsive enhancer
EST	expressed sequence tag
FFEM	freeze-fracture electron microscopy
GA	gibberellic acid
GSH	glutathione (reduced form)
GSSG	glutathione (oxidized form)
GST	glutathione-S-transferase

GUS	β-glucuronidase
HA	hydroxylamine
HL	high light
HR	hypersensitive response
HRGP	hydroxyproline-rich glycoprotein
hs	heat shock
HSE	heat shock promoter element
HSF	heat shock factor
HSP	heat shock protein
HU	hydroxylurea
IP_3	inositol-3-phosphate
JA	jasmonic acid
LEA	late embryogenesis abundant
LL	low light
LPC	leaf protein concentrate
LRO	light-responsive unit
LTP	lipid transfer protein
LTR	low temperature-responsive
LTRE	low temperature-responsive element
MAP	mitogen-activated protein
MAPK	MAP kinase
MGDG	monogalactosyldiacylglycerol
ODC	ornithine decarboxylase
osm	osmolal
PAGE	polyacrylamide gel electrophoresis
PAL	phenylalanine ammonia-lyase
PAP	3′-phosphoadenosine 5′-phosphate
PAPS	3′-phosphoadenosine 5′-phosphosulphate
PAR	photosynthetically active radiation
PC	phosphatidylcholine
PCD	programmed cell death
PCR	polymerase chain reaction
PDTC	pyrrolidine dithiocarbamate
PG	phosphatidylglycerol
PHGPX	phospholipid hydroperoxide glutathione peroxidase
PIP_2	phosphatidylinositol-4-phosphate-5-kinase
PIP5K	phosphatidylinositol-4,5-bisphosphate
PLC	phospholipase C
PPD	post-harvest physiological deterioration
PQ	plastoquinone
PR	pathogenesis-related
PSII	photosystem II
RLU	relative light unit
ROI	reactive oxygen intermediate

ROS	reactive oxygen species
Rubisco	ribulose-1,5-bisphosphate carboxylase/oxygenase
RUBP	ribulose diphosphate carboxylase
RWC	relative water content
SA	salicylic acid
SAM	S-adenosyl-methionine
SAR	systemic acquired resistance
SDS	sodium dodecyl sulphate
SOD	superoxide dismutase
SQDG	sulfoquinovosyldiacylglycerol
STRE	stress-responsive element
STS	stilbene synthase
TMV	tobacco mosaic virus

Preface

This volume arises from the Plant Protein Club's first Annual Symposium 'Plant proteins in abiotic stress responses: function, gene regulation and industrial applications' held in September 1998 at the University of York, U.K. This meeting brought together internationally respected academics investigating fundamental aspects of plant survival of abiotic stress with industrialists who are in a position to exploit their research commercially.

Abiotic environmental stresses which limit plant distribution and productivity include low and high temperature, salinity and water deficit. Over the last century human activities have increased the level of environmental stress in the form of pollutants such as ozone and heavy metals, levels of UV light reaching the biosphere and salinity in irrigated areas. Plants are sessile and therefore have developed mechanisms to survive environmentally extreme environments sometimes in vegetative stages of their life cycle. The molecular basis of these survival mechanisms is not only intrinsically interesting but may form a basis for enhanced plant productivity in currently marginal areas or increased latitude or season of cultivation of agronomically important crops.

The volume starts with a chapter describing recent developments in our understanding of the role of calcium as a second messenger in plants responding to diverse abiotic stresses. This is followed by two chapters looking at signal transduction and responses to light stress in the ultraviolet range. A recurring theme in plant responses to diverse environmental stresses is the production of reactive oxygen species and therefore a section has been devoted to the role of oxidative stress in abiotic stress responses. The book then moves on to address specific environmental stresses such as temperature, water deficit and salinity. These sections describe recent progress in our understanding of regulation of gene expression by these individual stresses as well as the biochemical function of individual proteins in the tolerance of stress. Throughout the book the relevance of research to industrial applications is emphasized.

M.F. Smallwood
C.M. Calvert
D.J. Bowles

Chapter 1

Calcium signalling in plants responding to stress

Heather Knight and Marc R. Knight

1. Background

The subject of calcium signalling in response to abiotic stresses in plants is vast, and this minireview will focus solely on the mediation of changes in intracellular calcium concentrations ($[Ca^{2+}]_i$) by abiotic stress.

2. Low temperature stress

A reduction in temperature causes an instantaneous influx of Ca^{2+} ions into the plant cell and consequent rise in $[Ca^{2+}]_i$. The link between cold treatment and calcium influx has been observed for some time. Increased influx of radiolabelled calcium ($^{45}Ca^{2+}$) into roots was observed in winter wheat treated at 2°C at high calcium concentrations (Erlandson and Jensen, 1989) and similar observations have been made in maize roots (Rincon and Hanson, 1986). The cold-induced influx of Ca^{2+} was found to be due to the opening of calcium channels rather than a general, cold-induced increased membrane permeability, and this has been demonstrated by the use of specific pharmacological agents. Influx of $^{45}Ca^{2+}$ into cold-shocked maize roots was strongly inhibited by the calcium channel blocker lanthanum, though only poorly by verapamil (Rincon and Hanson, 1986). Other workers also reported the inhibition of cold-induced calcium influx in maize roots by calcium channel blockers (De Nisi and Zocchi, 1996) with strong inhibition by verapamil and nifedipine, this particular study suggesting that voltage-gated calcium channels were involved. Reducing the temperature from 25 to 4°C caused a large increase in the influx of radiolabelled calcium ($^{45}Ca^{2+}$) in alfalfa protoplasts (Monroy and Dhindsa, 1995). The calcium channel blockers verapamil, lanthanum and nitrendipine or the calcium chelator BAPTA inhibited calcium influx (Monroy and Dhindsa, 1995). The exact nature of the cold-sensitive calcium channel(s) still remains unclear.

The cold-induced calcium influx via calcium channels results in an immediate transient elevation in cytosolic free calcium concentration ($[Ca^{2+}]_{cyt}$). This was first measured in intact whole plants using the recombinant aequorin approach (Knight *et al.*, 1991). Cold-induced $[Ca^{2+}]_{cyt}$ elevations have been measured in tobacco (Knight *et al.*,

Plant Responses to Environmental Stress, edited by M.F. Smallwood, C.M. Calvert and D.J. Bowles.
© 1999 BIOS Scientific Publishers, Oxford.

1991, 1992, 1993), *Arabidopsis* (Knight *et al.*, 1996; Lewis *et al.*, 1997; Polisensky and Braam, 1996) and the moss *Physcomitrella patens* (Russell *et al.*, 1996) using the recombinant aequorin method. The magnitude of the $[Ca^{2+}]_{cyt}$ elevation was markedly reduced by the calcium channel blocker lanthanum (Knight *et al.*, 1992, 1996), again indicating that it is due, at least in part, to the influx of calcium via plasma membrane calcium channels. Ding and Pickard (1993) identified a plasmalemmal mechanosensitive calcium channel, the activity of which was increased by a reduction in temperature, suggesting the involvement of this channel in the cold-induced elevation of $[Ca^{2+}]_i$. The authors speculated that this channel could be involved in the sensing of low temperature.

Work using aequorin has indicated that in addition to the influx of external calcium there is also a release of calcium from the vacuole, possibly via IP_3-sensitive channels (Knight *et al.*, 1996). *Arabidopsis* plants expressing recombinant aequorin targeted to the cytosolic face of the tonoplast membrane (an area we have termed the vacuolar microdomain), were used to demonstrate the $[Ca^{2+}]_{cyt}$ elevations occurring in this locale of the cell. Comparison of the cold-induced Ca^{2+} kinetics in this area with those throughout the cytosol (measured using aequorin targeted to the cytosol), suggested the involvement of vacuolar calcium in the cold shock response. Whether the influx component and the vacuolar release component of the $[Ca^{2+}]_{cyt}$ elevation have different roles is not yet known. The inhibition of the $[Ca^{2+}]$ elevation in the vacuolar microdomain by lithium chloride and neomycin, both inhibitors of phosphatidyl inositol cycling, suggested that IP_3 may be involved in mediating this release of calcium (Knight *et al.*, 1996).

3. Osmotic stress, drought and salinity stress

There is significant evidence to support the idea that Ca^{2+} release from intracellular stores plays a role in the responses both to salinity and osmotic stress or drought. The Ca^{2+}-sensitive fluorescent dye indo-1 was used to demonstrate directly that salinity causes an immediate increase in intracellular calcium concentration in maize root protoplasts (Lynch *et al.*, 1989), and this elevation of $[Ca^{2+}]_{cyt}$ was reduced by pretreatment with lithium. Results obtained with aequorin-expressing *Arabidopsis* support these data (see below). Together, these data suggested a role for calcium signalling, probably involving intracellular and extracellular sources of Ca^{2+}, and a role for inositol phosphates in the response to salinity.

In *Chara*, calcium influx increased in response to salinity, measured with $^{45}Ca^{2+}$ (Reid *et al.*, 1993). This effect appeared to be due to increased turgor, and was not Na^+-specific (Reid *et al.*, 1993). Using radiolabelled Ca^{2+}, it was shown that salt stress caused an increase in external calcium influx in the unicellular alga *Duniella salina* and that the magnitude of this influx depended on the degree of salt stress (Ko and Lee, 1995).

Further reports of $[Ca^{2+}]_{cyt}$ elevations in response to salinity included the demonstration of salt-induced calcium elevations in barley root protoplasts and in wheat aleurone cells loaded with the Ca^{2+}-sensitive fluorescent dyes indo-1 and fluo-3, respectively (Bittisnich *et al.*, 1989; Bush, 1996). Data obtained from experiments with *Arabidopsis* roots loaded with the ratiometric calcium sensitive dye, fura-2, indicated that in this tissue, NaCl treatment caused a decrease in cytosolic calcium levels (Cramer and Jones, 1996). These authors attributed this contrasting result to the fact that intact roots were used rather than isolated protoplasts as used in previous salinity studies (Bittisnich *et al.*, 1989; Lynch *et al.*, 1989). However, experiments using intact *Arabidopsis* seedlings (see below) showed increases in $[Ca^{2+}]_{cyt}$ (Knight *et al.*, 1997). The problems of variations in response from cell to cell which were encountered in these experiments may also contribute to these apparently contrasting results (Cramer and Jones, 1996).

$[Ca^{2+}]_{cyt}$ elevations in response to NaCl were measured in whole intact tobacco seedlings using the recombinant aequorin method (Knight *et al.*, 1991). Further investigations using whole *Arabidopsis* seedlings revealed that under the conditions used, very similar calcium transients resulted from salinity treatment or treatment with an osmotically equivalent dose of mannitol at two different concentrations of salt or mannitol (Knight *et al.*, 1997). Measurements made using plants expressing aequorin targeted to the tonoplast membrane (Knight *et al.*, 1996) indicated that the responses to salt or mannitol treatment involved a significant contribution of Ca^{2+} released from the vacuole (Knight *et al.*, 1997).

The effect of inhibitors of phosphoinositide cycling, lithium, U-73122 and neomycin, showed that this vacuolar-released calcium was likely to be triggered by IP_3-mediated release (Knight *et al.*, 1997), in agreement with earlier studies which suggested intracellular release of Ca^{2+} (Bittisnich *et al.*, 1989; Lynch and Lauchli, 1988; Lynch *et al.*, 1989).

4. Oxidative stress

The effect of the pollutant gas ozone on calcium transport was studied using vesicles from pinto bean leaves (Castillo and Heath, 1990). Ozone caused an increase in Ca^{2+} transport to the cytoplasmic side of the vesicles, the magnitude of the influx being dependent on the kinetics of the ozone exposure (Castillo and Heath, 1990).

The effects of oxidative stress on guard cells of *Commelina communis* were investigated by application of H_2O_2 (which generates hydroxyl radicals directly) and methyl viologen (which generates hydroxyl radicals as a secondary active oxygen species (AOS) after the formation of superoxide radicals) (McAinsh *et al.*, 1996). Using fluorescence ratio photometry of the calcium-sensitive dye fura-2, the authors measured changes in guard cell $[Ca^{2+}]_{cyt}$ in response to these two chemicals. Both methyl viologen (the herbicide paraquat) and H_2O_2 inhibited stomatal opening and promoted stomatal closure. Ethylene glycol bis-(β-aminoethylether) N,N,N',N'-tetraacetic acid (EGTA) completely abolished the effect of both methyl viologen and H_2O_2 on the opening and closing processes, when these chemicals were applied at concentrations of less than 10^{-5} M methyl viologen or H_2O_2 (McAinsh *et al.*, 1996).

The effect of oxidative stress on Ca^{2+} signalling in whole plants was studied using tobacco seedlings expressing recombinant aequorin (Price *et al.*, 1994). A rapid transient elevation of $[Ca^{2+}]_{cyt}$ was elicited by treatment with 10 mM H_2O_2. This elevation could be inhibited by lanthanum and by ruthenium red, which inhibits release from internal calcium stores, indicating that both intracellular and extracellular calcium stores contribute to the calcium elevation (Price *et al.*, 1994). The authors observed a refractory period during which further stimulation would not elicit a response. The recovery period had a duration of approximately 4–8 h, with recovery beginning after 30 min (Price *et al.*, 1994).

To investigate the mechanism of hydrogen peroxide perception, several inhibitors of the pathways surrounding AOS formation were used. Buthionine sulphoxime (BSO) is an inhibitor of glutathione (GSH) synthesis and hydroxylurea (HU) and hydroxylamine (HA) inhibit ascorbate peroxidase, the first enzyme of the Halliwell–Asada pathway which degrades hydrogen peroxide. Both groups of chemicals should retard the rate of hydrogen peroxide metabolism, allowing a longer period over which hydrogen peroxide can cause $[Ca^{2+}]_{cyt}$ elevations. However, the effect of BSO on the pro-oxidant/antioxidant ratio is the opposite of that caused by HA or HU. (The primary pro-oxidants referred to here are dehydroascorbic acid, quinones and GSH and the antioxidants are the equivalent reduced forms.) BSO (which should elevate the pro-oxidant/antioxidant ratio) caused increase in the peak height of the hydrogen peroxide-induced Ca^{2+} response whereas HU and HA (which would be predicted to lower the pro-oxidant/antioxidant ratio) caused the Ca^{2+} response kinetic to become slower, with a lower peak value and a slower decline (Price *et al.*, 1994). These results indicated that the H_2O_2-induced $[Ca^{2+}]_{cyt}$ elevations occurred not in direct response to H_2O_2 itself but in response to the pro-oxidant/antioxidant ratio.

5. Heat stress

Heat-induced elevations of $[Ca^{2+}]_{cyt}$ were inferred by measurement of significantly enhanced levels of Ca^{2+} uptake by suspension-cultured pear cells and protoplasts after exposure to temperatures of 38°C (Klein and Ferguson, 1987). Direct measurement of heat shock-induced Ca^{2+} elevations has been performed in pea mesophyll protoplasts using the fluorescent dye indo-1 (Biyaseheva *et al.*, 1993). In these experiments heat shock induced a four-fold increase in $[Ca^{2+}]_{cyt}$. Further studies of heat shock-induced Ca^{2+} elevations have been performed in tobacco using the recombinant aequorin technique (Gong *et al.*, 1998). Heat shock of tobacco seedlings expressing recombinant aequorin revealed that after 5–35 min at elevated temperatures (39, 43 or 47°C) $[Ca^{2+}]_{cyt}$ was elevated. The elevation of $[Ca^{2+}]_{cyt}$ was prolonged but transient. Brief treatments with hot water did not elevate $[Ca^{2+}]_{cyt}$ (Gong *et al.*, 1998; Knight *et al.*, 1991). Preincubation with Ca^{2+} caused the heat-induced Ca^{2+} elevation to begin more quickly, and to reach a higher than normal peak level. In contrast, pretreatment with EGTA caused a reduced and slower response (Gong *et al.*, 1998). The response was attenuated by repeated heat stimulation, although plants were able during this period to respond to other stresses normally, such as cold. The magnitude of the signal was inhibited by the channel blockers ruthenium red and lanthanum as well as by the phospholipase C inhibitor, neomycin. These data suggested the possibility that both extracellular and intracellular (possibly IP_3-mediated) stores are involved in contributing to the $[Ca^{2+}]_{cyt}$ elevation.

6. Mechanical stress

Using the recombinant aequorin method, it has been shown that $[Ca^{2+}]_{cyt}$ is elevated by touch stimulation (Knight *et al.*, 1991). Such elevations have been demonstrated in tobacco (Knight *et al.*, 1991, 1992, 1993), *Arabidopsis* (Knight *et al.*, 1995; A. J. Wright and M. R. Knight, unpublished data) and the moss *Physcomitrella patens* (Russell *et al.*, 1996). When increasing wind forces were applied to tobacco seedlings, a graded response was observed, with increases in the magnitude of the $[Ca^{2+}]_{cyt}$ elevation occurring with increasing force (Knight *et al.*, 1992). The magnitude of the elevation correlated with the amount of time the seedling was in motion when responding to wind or touch and when seedlings were permanently bent no $[Ca^{2+}]_{cyt}$ elevation was measured (Knight *et al.*, 1992). This suggested that the $[Ca^{2+}]_{cyt}$ elevation is only elicited when tension and compression of cells continues to change (Knight *et al.*, 1992). In similar experiments, expulsion of a known volume of air on to a seedling caused a touch response, the magnitude of which increased with rapidity of expulsion (Haley *et al.*, 1995). The effect was also seen in individual cells, as demonstrated by injecting isotonic medium into a suspension of aequorin-expressing tobacco protoplasts (Haley *et al.*, 1995). The $[Ca^{2+}]_{cyt}$ response to mechanical stress appears to occur in all cells and has been observed in the aerial parts (A. J. Wright and M. R. Knight, unpublished data) and in the meristematic, elongation and differentiated zones and root cap of *Arabidopsis* roots (Legué *et al.*, 1997).

The subcellular origins of the Ca^{2+} signal are still not clear. Unlike the response to cold shock or osmotic stress, the $[Ca^{2+}]_{cyt}$ elevation elicited by touch stimulation cannot be inhibited by the plasma membrane channel blocker La^{3+} or by gadolinium, a stretch-activated calcium channel blocker (Yang and Sachs, 1989) but rather is inhibited by ruthenium red, a putative inhibitor of mitochondrial and endoplasmic reticulum calcium channels (Knight, 1992). A similar lack of effect of La^{3+} or Gd^{3+} was seen in tobacco protoplasts (Haley *et al.*, 1995). Ruthenium red did not, however, affect the touch-induced $[Ca^{2+}]_{cyt}$ elevation elicited in acid-treated epidermal strips in which the only viable cells were guard cells and trichomes (Haley *et al.*, 1995) and in the filamentous moss *Physcomitrella patens*, EGTA and La^{3+} did cause some reduction of the response, making the overall interpretation of these data difficult. Attenuation of the touch response by repeated stimulation did not prevent the subsequent elicitation of a normal Ca^{2+} response to cold stimulation, again indicating that Ca^{2+} stores other than the vacuole and the apoplast are involved (Knight *et al.*, 1992). Experiments with plants expressing subcellular targeted aequorin have indicated that the intracellular store involved is unlikely to be the vacuole or the chloroplast (H. Knight, unpublished data), and the more likely nucleus, mitochondria and endoplasmic reticulum have yet to be explored.

7. Key questions

The key question relating to calcium signalling is that of specificity. How can calcium be involved in the signalling pathways mediating the responses to so many primary stimuli, and yet those pathways manage to be specific to the given stimulus? Related to this is the question of how information is encoded in the form of changes of intracellular calcium, that is what information is encoded by a calcium 'signature' generated by a given stimulus? Is a calcium signal simply an 'on/off' switch that needs to be activated, with specificity of signalling actually encoded elsewhere in the signalling pathways? Or does the magnitude, duration and sometime complex nature, for example oscillations of calcium signatures, encode specific information?

8. Bottlenecks hindering progress

The key to answering the questions above would rely on being able to generate complex patterns of intracellular calcium elevations *in the absence of a primary signal*. If it were possible to create calcium signals 'to order' then the effect of changing the parameters, for example magnitude, period, oscillation rate and so on, of these calcium signals could be gauged in terms of their effects downstream, for example specific gene expression, guard cell closure and so on. By studying the pattern of end responses generated by changing patterns of calcium signature, it could be determined if and how calcium signalling can mediate specificity. This type of work is, however, currently technically very difficult. Overcoming this particular bottleneck would be a significant achievement. The use of inhibitors, for example calcium channel blockers, has yielded important information on calcium signalling, but because the specificity of their action is relatively unknown, especially in plants, this has helped little in terms of understanding specificity in calcium signalling. A more targeted and specific approach needs to be developed. This could be in the form of recombinant versions of agonists/antagonists and the identification of calcium signalling mutants, namely a combination of molecular biology and molecular genetics to tackle this thorny problem. Apart from this consideration of the temporal aspects of calcium signalling, it becomes increasingly important that knowledge is gained about the spatial aspects of calcium signalling, both at the subcellular level and also at the whole tissue and plant level. Mapping changes in intracellular calcium concentration as well as calcium-sensitive components of signalling, for example calmodulin and calcium-dependent protein kinases and downstream responses, for example stress-induced gene expression, will be vital. The three stages of calcium signalling need to be investigated at the subcellular and whole plant level, and if possible in 3D!

9. Future directions

Future research should be directed at the key questions of specificity in calcium signalling as described above. Understanding how the integration of multiple signals may be occurring at the calcium signalling level is also very important and should be examined. Identifying the key calcium-dependent signalling components downstream of intracellular calcium changes needs to be done. Understanding how specificity is encoded at this level is likely to be important, especially if some (or all!) of specificity cannot be explained by information encoded in calcium signature alone. Highly targeted molecular and genetic approaches to these problems need to be applied to tackle all these problems.

References

Bittisnich, D., Robison, D. and Whitecross, M. (1989) In: *Plant Membrane Transport: The Current Position. Proceedings of the Eighth International Workshop on Plant Membrane Transport* (eds J. Dainty, M.I. de Michelis, E. Marre and F. Rasi-Caldogno). Elsevier Science Publishing Company, Venice, pp. 681–682.

Biyaseheva, A.E., Molotkovskii, Y.G. and Mamonov, L.K. (1993) Increase of free Ca^{2+} in the cytosol of plant protoplasts in response to heat stress as related to Ca^{2+} homeostasis. *Russian Plant Physiol.* (English translation) **40:** 540–544.

Bush, D.S. (1996) Effects of gibberellic acid and environmental factors on cytosolic calcium in wheat aleurone cells. *Planta* **199:** 89–99.

Castillo, F.J. and Heath, R.L. (1990) Calcium ion transport in membrane vesicles from pinto bean leaves and its alteration after ozone exposure. *Plant Physiol.* **94:** 788–795.

Cramer, G.R. and Jones, R.L. (1996) Osmotic stress and abscisic acid reduce cytosolic calcium activities in roots of *Arabidopsis thaliana. Plant Cell Environ.* **19:** 1291–1298.

De Nisi, P. and Zocchi, G. (1996) The role of calcium in the cold shock responses. *Plant Sci.* **121:** 161–166.

Ding, J.P. and Pickard, B.G. (1993) Modulation of mechanosensitive calcium-selective cation channels by temperature. *Plant J.* **3:** 713–720.

Erlandson, A.G.I. and Jensen, P. (1989) Influence of low temperature on regulation of rubidium and calcium influx in roots of winter wheat. *Physiol. Plant.* **75:** 114–120.

Gong, M., van der Luit, A.H., Knight, M.R. and Trewavas, A.J. (1998) Heat-shock-induced changes in intracellular Ca^{2+} level in tobacco seedlings in relation to thermotolerance. *Plant Physiol.* **116:** 429–437.

Haley, A., Russell, A.J., Wood, N., Allan, A.C., Knight, M., Campbell, A.K. and Trewavas, A.J. (1995) Effects of mechanical signaling on plant cell cytosolic calcium. *Proc. Natl Acad. Sci. USA* **92:** 4124–4128.

Klein, J.D. and Ferguson, I.B. (1987) Effect of high temperature on calcium uptake by suspension-cultured pear fruit cells. *Plant Physiol.* **84:** 153–156.

Knight, H., Trewavas, A.J. and Knight, M.R. (1996) Cold calcium signaling in *Arabidopsis* involved two cellular pools and a change in calcium signature after acclimation. *Plant Cell* **8:** 489–503.

Knight, H., Trewavas, A.J. and Knight, M.R. (1997) Calcium signalling in *Arabidopsis thaliana* responding to drought and salinity. *Plant J.* **12:** 1067–1078.

Knight, M.R., Campbell, A.K., Smith, S.M. and Trewavas, A.J. (1991) Transgenic plant aequorin reports the effects of touch and cold-shock and elicitors on cytoplasmic calcium. *Nature* **352:** 524–526.

Knight, M.R., Smith, S.M. and Trewavas, A.J. (1992) Wind-induced plant motion immediately increases cytosolic calcium. *Proc. Natl Acad. Sci. USA* **89:** 4967–4971.

Knight, M.R., Read, N.D., Campbell, A.K. and Trewavas, A.J. (1993) Imaging calcium dynamics in living plants using semi-synthetic recombinant aequorins. *J. Cell Biol.* **121:** 83–90.

Knight, M.R., Knight, H. and Watkins, N.J. (1995) Calcium and the generation of plant form. *Phil. Trans. Roy. Soc. Lond. B Biol. Sci.* **350:** 83–86.

Ko, J.H. and Lee, S.H. (1995) Role of calcium in the osmoregulation under salt stress in *Dunaliella salina. J. Plant Biol.* **38:** 243–250.

Legué, V., Blancaflor, E., Wymer, C., Perbal, G., Fantin, G. and Gilroy, S. (1997) Cytoplasmic free Ca^{2+} in *Arabidopsis* roots changes in response to touch but not gravity. *Plant Physiol.* **114:** 789–800.

Lewis, B.D., Karlin-Neumann G., Davis, R.W. and Spalding, E.P. (1997) Ca^{2+}-activated anion channels and membrane depolarizations induced by blue light and cold in *Arabidopsis* seedlings. *Plant Physiol.* **114:** 1327–1334.

Lynch, J. and Lauchli, A. (1988) Salinity affects intracellular calcium in corn root protoplasts. *Plant Physiol.* **87:** 351–356.

Lynch, J., Polito, V.S. and Lauchli, A. (1989) Salinity stress increases cytoplasmic calcium activity in maize root protoplasts. *Plant Physiol.* **90:** 1271–1274.

McAinsh, M.R., Clayton, H., Mansfield, T.A. and Hetherington, A.M. (1996) Changes in stomatal behavior and guard cell cytosolic free calcium in response to oxidative stress. *Plant Physiol.* **111:** 1031–1042.

Monroy, A.F. and Dhindsa, R.S. (1995) Low-temperature signal transduction: induction of cold acclimation-specific genes of alfalfa by calcium at 25 degrees C. *Plant Cell* **7:** 321–331.

Polisensky, D.H. and Braam, J. (1996) Cold-shock regulation of the *Arabidopsis* TCH genes and the effects modulating intracellular calcium levels. *Plant Physiol.* **111:** 1271–1279.

Price, A.H., Taylor, A., Ripley, S.J., Griffiths, A., Trewavas, A.J. and Knight, M.R. (1994) Oxidative signals in tobacco increase cytosolic calcium. *Plant Cell* **6:** 1301–1310.

Reid, R.J., Tester, M. and Smith, F.A. (1993) Effects of salinity and turgor on calcium influx in *Chara*. *Plant Cell Environ.* **16:** 547–554.

Rincon, M. and Hanson, J.B. (1986) Controls on calcium ion fluxes in injured or shocked corn (*Zea mays*) root cells: importance of proton pumping and cell membrane potential. *Physiol. Plant.* **67:** 576–583.

Russell, A.J., Knight, M.R., Cove, D.J., Knight, C.D., Trewavas, A.J. and Wang, T.L. (1996) The moss, *Physcomitrella patens*, transformed with apoaequorin cDNA responds to cold shock, mechanical perturbation and pH with transient increases in cytoplasmic calcium. *Transgenic Res.* **5:** 167–170.

Yang, X.C. and Sachs, F. (1989) Block of stretch-activated ion channels in *Xenopus* oocytes by gadolinium and calcium ions. *Science* **243:** 1068–1071.

Chapter 2

Regulation of phenylpropanoid and flavonoid biosynthesis genes by UV-B in *Arabidopsis*

Gareth I. Jenkins

1. Background

Genes encoding enzymes of the phenylpropanoid and flavonoid biosynthesis pathways are stimulated when plants are exposed to a range of abiotic and biotic stresses (Dixon and Paiva, 1995). The reason for this is that many of the products of these branching pathways are important in plant protection. For instance, it is well established that several compounds derived from the phenylpropanoid pathway have antipathogenic properties (Dixon and Paiva, 1995). In addition, some flavonoids and hydroxycinnamates absorb Ultraviolet-B (UV-B) (280–320 nm) and function as sun screens in the epidermal layers (Jenkins *et al.*, 1997; Tevini and Teramura, 1989). The reason that anthocyanins and various other flavonoids and phenylpropanoids accumulate under stress conditions is less clear, although it is quite possible that they function as antioxidants.

The fact that genes encoding components of these pathways are stimulated by numerous environmental stimuli makes them an excellent system for dissecting the corresponding signal transduction pathways. Most of this research has focused on genes encoding phenylalanine ammonia-lyase (PAL) and chalcone synthase (CHS), the first enzymes of the phenylpropanoid and flavonoid biosynthesis pathways. Research in my laboratory is concerned with the abiotic stress signalling pathways and, in particular, the mechanisms through which UV-B regulates *CHS* and *PAL* expression. The research is focused on *Arabidopsis* because it allows a powerful combination of molecular, genetic, biochemical and cell physiological approaches to be employed.

The key questions in this research are: How is UV-B radiation initially perceived by the plant? What are the components of the signal transduction pathways that couple this initial perception to the regulation of transcription of *PAL* and *CHS* genes? How does UV-B signal transduction integrate into the network of signalling pathways that regulate these genes?

We are attempting to identify UV-B stimulus perception and transduction components using an *Arabidopsis* cell suspension system. An advantage of this system is that it permits both a pharmacological approach and direct measurements of signal transduction processes. In many respects it resembles the cell culture systems that have been

---

so successfully employed in studies of animal signal transduction. In addition, we are using *Arabidopsis* to isolate mutants altered specifically in the UV-B induction of *CHS* expression. This will permit the functional characterization of signalling and effector components and the isolation of their genes. It is likely that the biochemical/cell physiological and molecular genetic approaches will be complementary in terms of the components they identify. Furthermore, we are attempting to integrate these approaches because this provides a powerful means of dissecting UV-B signal transduction.

2.   How is UV-B perceived?

UV-B is reported to affect the expression of several genes in plants, apart from *PAL* and *CHS* (Conconi *et al.*, 1996; Green and Fluhr, 1995; Mackerness *et al.*, 1997), and also to have morphogenetic effects (Kim *et al.*, 1998; Tevini and Teramura, 1989). In animal cells, UV-induced DNA damage provides a signal that stimulates gene expression, but there is no direct evidence for an equivalent mechanism in plants. However, it appears that UV-B may induce gene expression because it causes oxidative stress. Green and Fluhr (1995) reported that the generation of reactive oxygen species (ROS) stimulated expression of the PR1 gene in tobacco. Furthermore, the induction of this gene following UV-B exposure was prevented by the application of antioxidant compounds. This led the authors to propose that the production of ROS during UV-B exposure mediated PR1 expression. Our experiments in *Arabidopsis* (J.C. Long and G.I. Jenkins, unpublished data) indicate that treatments that generate oxidative stress do not induce *CHS* expression and that the UV-B induction of *CHS* is not prevented by antioxidant compounds. Thus it appears that ROS do not mediate this particular UV-B response. Conconi *et al.* (1996) reported that the induction of proteinase inhibitor I and II genes in tomato by UV-C and UV-B was absent in a mutant defective in octadecanoid signalling, which produces jasmonic acid from linolenic acid. Hence, the authors concluded that the response to UV-B was mediated by this signalling pathway. However, in *Arabidopsis* cells we have found that the induction of *PAL* by jasmonate occurs with slower kinetics than UV-B induction and, moreover, the two pathways have different pharmacological properties (S. Grundy, V. Meier and G.I. Jenkins, unpublished data). Thus, in the case of *PAL*, the UV-B induction does not appear to be mediated by jasmonate. It is therefore likely that the effects of UV-B on gene expression in plants are mediated by several distinct pathways.

   An alternative possible mechanism for UV-B perception is via a photoreceptor. Phytochrome absorbs UV-B radiation and it is possible that it may mediate, or at least be required for, some UV-B effects (Kim *et al.*, 1998). However, the phytochrome induction of *CHS* in *Arabidopsis* is confined to very young seedlings (Kaiser *et al.*, 1995), and so does not account for UV-B-induced expression in older leaves. Moreover, phytochrome deficient *Arabidopsis* mutants retain the UV/blue light induction of *CHS* expression (Batschauer *et al.*, 1996). It is possible that plants have one or more specific UV-B photoreceptors (Jenkins *et al.*, 1997). If so, there could be a similarity to the cryptochromes that detect UV-A and blue light (Ahmad and Cashmore, 1996). The cryptochromes have flavin and pterin chromophores which could conceivably function in detecting UV-B (Jenkins *et al.*, 1997). Studies with cryptochrome-deficient mutants indicate that these photoreceptors are not involved in UV-B perception, at least in the induction of *CHS* (Fuglevand *et al.*, 1996; W.J. Valentine and G.I. Jenkins, unpublished data). Thus, given that there is no direct evidence for a UV-B photoreceptor, the mechanism of UV-B perception in *CHS* regulation remains unknown.

3. UV-B signal transduction regulating *PAL* and *CHS* in *Arabidopsis*

We have used an *Arabidopsis* cell suspension culture to investigate the signal transduction processes concerned with the induction of *CHS* expression by UV-B light. *CHS* transcript levels in these cells are induced by UV-B and UV-A/blue light, but not red or far-red light, which is the same as in mature *Arabidopsis* leaves (Christie and Jenkins, 1996). Following illumination with UV-B, *CHS* transcripts accumulate in the cells within 4 h. We examined the effects of various inhibitors of signal transduction components on this inductive response, using appropriate controls to show that their effects were specific (Christie and Jenkins, 1996).

The UV-B induction of *CHS* transcripts in the cells was strongly inhibited by both nifedipine, which inhibits voltage-gated calcium channels, and ruthenium red, which inhibits calcium channels in vacuolar membranes (Allen *et al.*, 1995) and probably other membranes (Marshall *et al.*, 1994). However, lanthanum, which competes externally with calcium for uptake into cells, and verapamil, another calcium channel blocker, did not inhibit. Nifedipine and ruthenium red do not have general inhibitory effects on the cells because they do not affect the induction of control transcripts. Moreover, lanthanum and verapamil do inhibit other gene expression responses in the cells (J.M. Christie, V. Maier and G.I. Jenkins, unpublished data). The lanthanum and ruthenium red data suggest that the UV-B induction of *CHS* may require calcium flux across an internal membrane and a specific pool of cellular calcium may therefore be involved.

Attempts to induce *CHS* expression in the *Arabidopsis* cells by artificially elevating cytosolic calcium with the ionophore A23187 and 10 mM external calcium were unsuccessful (Christie and Jenkins, 1996). However, the ionophore treatment did raise the cytosolic calcium concentration, because transcripts of the *Arabidopsis TCH3* (*TOUCH3*) gene, which is induced by cytosolic calcium, increased. If the cytosolic calcium concentration is raised from intracellular stores, the failure to induce expression by artificially elevating cytosolic calcium could indicate that UV-B illumination produces an additional essential signal: that is, the increase in calcium is necessary but not sufficient for the response. An alternative possibility is that UV-B promotes a localized increase in calcium, for example in a microdomain at a particular intracellular membrane or in an organelle, and that this pool is not raised by the ionophore treatment.

Recent experiments (Long and Jenkins, 1998) indicate that UV-B initiates electron transport processes in the plasma membrane and these may be required in conjunction with the calcium signal to stimulate *CHS* expression. The evidence that electron transport is involved is that the cell impermeant electron acceptor ferricyanide and the flavoprotein antagonist diphenylene iodonium both strongly inhibit *CHS* expression, whereas they do not inhibit control gene expression responses.

Long and Jenkins (1998) further observed that UV-B illumination appeared to promote an efflux of calcium from the cytosol under certain conditions. *TCH3* transcripts increased in response to calcium ionophore treatment in a low, non-inductive fluence rate of white light, but this response was greatly reduced in UV-B. This suggests that UV-B treatment lowered the cytosolic calcium concentration established by the ionophore. A similar result was observed in UV-A/blue light, but not red light. The inhibition of ionophore-induced *TCH3* expression by UV-B light (but not UV-A/blue light) was overcome by preincubation of the cells with the calmodulin antagonist W-7. A possible explanation of these results is that UV-B promotes calcium efflux from the cytosol via a calmodulin-dependent $Ca^{2+}ATPase$.

Thus, our working hypothesis is that UV-B perception initiates plasma membrane electron transport, which generates a signal that regulates calcium flux into and out of an internal store. The calcium concentration in the cytosol adjacent to the internal store is proposed to be an essential component of the signal that induces *CHS* expression.

The pharmacological experiments with the *Arabidopsis* cell culture show that the UV-B light signalling pathway is distinct from both the UV-A/blue light and phytochrome signalling pathways regulating *CHS* expression (Christie and Jenkins, 1996; Jenkins, 1997; Long and Jenkins, 1998). This conclusion is consistent with genetic and photobiological evidence, referred to above, showing that the UV-B induction of *CHS* is not mediated by cryptochromes or phytochromes.

It is important to establish how UV-B signal transduction causes the stimulation of *CHS* transcription. As a first step in this research we have identified the promoter elements of the gene that are required for the UV-B response. Hartmann *et al.* (1998) reported that the *Arabidopsis CHS* promoter contains a 41 bp light-responsive unit (LRU) with a high degree of homology to equivalent regions in other *CHS* promoters. The *Arabidopsis CHS* LRU is sufficient to confer strong UV-B light-induction of the β-glucuronidase (GUS) reporter in transient expression experiments with protoplasts isolated from the *Arabidopsis* cell culture. The LRU contains sequence elements that recognize basic leucine zipper and myb classes of transcription factors. However, *Arabidopsis* contains numerous factors of these types encoded in multigene families. Hence, the challenge now is to identify the specific transcription factors that interact with the *CHS* LRU in *Arabidopsis* and to understand how their biogenesis and/or activity is regulated by UV-B light.

Experiments with specific inhibitors indicated that serine/threonine protein kinase activity and protein phosphatase activities are required for the UV-B light-induction of *CHS* in the *Arabidopsis* cells (Christie and Jenkins, 1996). There is evidence that phosphorylation is involved in the activation of the bZIP transcription factors that associate with the LRU1 in parsley (Harter *et al.*, 1994), providing a link to the signalling pathways. In addition, *CHS* induction by UV-B was prevented by cycloheximide in *Arabidopsis* cells (Christie and Jenkins, 1996), demonstrating a requirement for protein synthesis. Thus, one or more signalling components or effectors, possibly a transcription factor, needs to be synthesized to effect the *CHS* transcription response. The identification of the relevant component(s) is therefore a priority.

Hence some insights are being gained into the cellular mechanisms through which

UV-B is perceived and signal transduction is coupled to *CHS* gene transcription. However, pharmacological studies clearly have their limitations and it is therefore necessary to identify directly the signal transduction processes initiated by UV-B light. This will require the application of cell physiological approaches in cells and plants.

4. Interactions between UV-B, UV-A and blue light signalling pathways

Induction by UV-B alone is insufficient to account for the level of *CHS* transcription observed in mature *Arabidopsis* leaves in natural daylight. The reason for this is that the response to UV-B is enhanced by synergistic interactions with UV-A and blue light signalling pathways which serve to maximize the level of *CHS* expression (Fuglevand *et al.*, 1996). The level of *CHS-GUS* expression in transgenic *Arabidopsis* exposed to UV-B and blue light together was four- to eight-fold greater than in plants given either light quality alone. A similar synergistic interaction was observed between UV-B and UV-A light, but not with UV-A and blue light, which gave an additive effect.

Further experiments revealed that blue light given before UV-B resulted in a synergistic increase in the level of *CHS-GUS* expression and *CHS* transcript accumulation, whereas UV-B given before blue light did not. Thus blue light produced a 'signal' that enhanced the subsequent response to UV-B. Moreover, the enhancement was still observed when a several-hour dark period was introduced between the blue and UV-B treatments, indicating that the signal derived from blue light was relatively stable. These results are similar to those reported previously for *CHS* expression in parsley cell cultures (Ohl *et al.*, 1989). Fuglevand *et al.* (1996) additionally reported that the UV-A 'signal' was transient, not stable, because a synergistic interaction between UV-A and UV-B was observed only when the two treatments were given simultaneously. This difference in the stability of the signals derived from UV-A and blue light indicates that the phototransduction pathways that produce them are distinct. In support of this hypothesis, illumination of plants with blue light followed by UV-A and UV-B together resulted in approximately twice the level of *CHS-GUS* expression observed with either synergistic combination (UV-B plus blue light or UV-B plus UV-A light). Thus, there appears to be an additive effect of two synergistic interactions, which indicates the remarkable complexity involved in interactions between signalling pathways.

It is evident that the UV-B, UV-A and blue light signal transduction pathways are only a subset of the signalling pathways that regulate *CHS* and *PAL* genes. We have found that a number of kinetically and pharmacologically distinct pathways stimulate these genes in *Arabidopsis*. Furthermore, there is evidence of additional synergistic interactions and of negative regulation (Lozoya *et al.*, 1991) between particular pathways. Thus, it is important to think in terms of signal transduction networks rather than isolated pathways. Interactions or 'cross talk' between components of this network achieve the integration of responses to multiple environmental and endogenous stimuli (Jenkins, 1998).

5. Mutants altered in the UV-B induction of *CHS*

We have used a genetic approach to identify components involved in the regulation of

CHS expression by UV-B. Mutant isolation and characterization provides a powerful means of identifying and defining the functions of signalling components and downstream effectors regulating gene expression. Moreover, it provides a means of isolating the corresponding wild-type genes.

Jackson *et al.* (1995) used a transgene expression screen to isolate mutants altered specifically in the light-regulation of *CHS* gene expression. M2 seedlings derived from a transgenic line expressing *CHS-GUS* were screened for altered GUS activity in the light. One mutant, *icx1* (increased chalcone synthase expression) had elevated light-induction of *CHS-GUS* expression and increased levels of transcripts of *CHS* and other flavonoid biosynthesis genes. Expression in darkness was unaltered, so the mutant has an enhanced response to light. The increase in *CHS* expression is present in UV-B and UV-A/blue light (H.K. Wade and G.I. Jenkins, unpublished data), so ICX1 appears to be a negative regulator of both signalling pathways. Furthermore, the *icx1* mutant has increased *CHS* expression in response to non-light stimuli (H.K. Wade and G.I. Jenkins, unpublished data) and has altered epidermal development (Jackson *et al.*, 1995; J.A. Jackson, B.A. Brown and G.I. Jenkins, unpublished data). Thus it appears that ICX1 is an important regulator of a number of epidermal processes.

Several additional mutants with altered *CHS* expression have since been isolated using the above screen (G. Fuglevand and G.I. Jenkins, unpublished data). One, *icx2*, shows strongly elevated *CHS-GUS* expression, *CHS* transcript levels and anthocyanin induction. Expression is increased in both UV-B and UV-A/blue light (M.R. Shenton and G.I. Jenkins, unpublished data). Hence this mutant defines a further negative regulator of *CHS* transcription.

A challenge for the future will be to clone the genes encoding the negative regulators identified genetically and to understand how they function in relation to the signal transduction pathways characterized by pharmacological experiments.

Acknowledgements

The author thanks the Biotechnology and Biological Sciences Research Council, the Gatsby Charitable Foundation and the British Council for their support of research in his laboratory.

References

Ahmad, M. and Cashmore, A.R. (1996) Seeing blue: the discovery of cryptochrome. *Plant Mol. Biol.* **30:** 851–861.

A.-H.-Mackerness, S., Jordan, B.R. and Thomas, B. (1997) UV-B effects on the expression of genes encoding proteins involved in photosynthesis. In: *Plants and UV-B: Responses to Environmental Change* (ed. P.J. Lumsden). Cambridge University Press, Cambridge, pp. 113–134.

Allen, G.J., Muir, S.R. and Sanders, D. (1995) Release of Ca^{2+} from individual plant vacuoles by both InsP$_3$ and cyclic ADP-ribose. *Science* **268:** 735–737.

Batschauer, A., Rocholl, M., Kaiser, T., Nagatani, A., Furuya, M. and Schäfer, E. (1996) Blue and UV-A light-regulated *CHS* expression in *Arabidopsis* independent of phytochrome A and phytochrome B. *Plant J.* **9:** 63–69.

Christie, J.M. and Jenkins, G.I. (1996) Distinct UV-B and UV-A/blue light signal transduction pathways induce chalcone synthase gene expression in Arabidopsis cells. *Plant Cell* **8:** 1555–1567.

Conconi, A., Smerdon, M.J., Howe, G.A. and Ryan, C.A. (1996) The octadecanoid signalling pathway in plants mediates a response to ultraviolet radiation. *Nature* **383:** 826–829.

Dixon, R.A. and Paiva, L.A. (1995) Stress-induced phenylpropanoid metabolism. *Plant Cell* **7:** 1085–1097.

Fuglevand, G., Jackson, J.A. and Jenkins, G.I. (1996) UV-B, UV-A and blue light signal transduction pathways interact synergistically to regulate chalcone synthase gene expression in Arabidopsis. *Plant Cell* **8:** 2347–2357.

Green, R. and Fluhr, R. (1995) UV-B-induced PR-1 accumulation is mediated by active oxygen species. *Plant Cell* **7:** 203–212.

Harter K., Kircher S., Frohnmeyer H., Krenz M., Nagy F. and Schäfer, E. (1994) Light-regulated modification and nuclear translocation of cytosolic G-box binding factors in parsley. *Plant Cell* **6:** 545–559.

Hartmann, U., Valentine, W.J., Christie, J.M., Hays, J., Jenkins, G.I. and Weisshaar, B. (1998) Identification of UV/blue light-response elements in the *Arabidopsis thaliana* chalcone synthase promoter using a homologous protoplast transient expression system. *Plant Mol. Biol.* **36:** 741–754.

Jackson, J.A., Fuglevand, G., Brown, B.A., Shaw, M.J. and Jenkins, G.I. (1995) Isolation of *Arabidopsis* mutants altered in the light-regulation of chalcone synthase gene expression using a transgenic screening approach. *Plant J.* **8:** 369–380.

Jenkins, G.I. (1997) UV and blue light signal transduction in Arabidopsis. *Plant Cell Environ.* **20:** 773–778.

Jenkins, G.I. (1998) Signal transduction networks and the integration of responses to environmental stimuli. *Adv. Bot. Res.* **29:** 55–73.

Jenkins, G.I., Fuglevand, G. and Christie, J.M. (1997) UV-B perception and signal transduction. In: *Plants and UV-B: Responses to Environmental Change* (ed. P.J. Lumsden). Cambridge University Press, Cambridge, pp. 135–156.

Kaiser, T., Emmler, K., Kretsch, T., Weisshaar, B., Schäfer, E. and Batschauer, A. (1995) Promoter elements of the mustard *CHS1* gene are sufficient for light-regulation in transgenic plants. *Plant Mol. Biol.* **28:** 219–229.

Kim, B.C., Tennessen, D.J. and Last, R.L. (1998) UV-B-induced photomorphogenesis in *Arabidopsis thaliana*. *Plant J.* **15:** 667–674.

Long, J.C. and Jenkins, G.I. (1998) Involvement of plasma membrane redox activity and calcium homeostasis in the UV-B and UV-A/blue light induction of gene expression in *Arabidopsis*. *Plant Cell* **10:** 2077–2086.

Lozoya, E., Block, A., Lois, R., Hahlbrock, K. and Scheel, D. (1991) Transcriptional repression of light-induced flavonoid synthesis by elicitor treatment of cultured parsley cells. *Plant J.* **1:** 227–234.

Marshall, J., Corzo A., Leigh, R.A. and Sanders, D. (1994) Membrane potential-dependent calcium transport in right-side-out plasma membrane vesicles from *Zea mays* L. roots. *Plant J.* **5:** 683–694.

Ohl, S., Hahlbrock, K. and Schäfer, E. (1989) A stable blue-light-derived signal modulates ultraviolet-light-induced activation of the chalcone synthase gene in cultured parsley cells. *Planta* **177:** 228–236.

Tevini, M. and Teramura, A.H. (1989) UV-B effects on terrestrial plants. *Photochem. Photobiol.* **50:** 479–487.

Chapter 3

Effects of UV-B radiation on plants: gene expression and signal transduction pathways

Soheila A.-H.-Mackerness and Brian Thomas

1. Background

Due to the depletion of the stratospheric ozone layer, through use and ultimate release of man-made ozone-destroying chemicals, the level of ultraviolet-B radiation (UV-B: 280–320 nm) reaching the Earth's surface is increasing (Kerr and McElroy, 1993; Madronich et al., 1995). Exposure of plants to UV-B radiation alone is highly detrimental to their growth and survival. Fortunately, with accompanying visible radiation, the effects are less severe, but UV-B is still damaging. The effects of UV-B on plants include reductions in photosynthesis resulting in decreased growth, anatomical and morphological changes and increases in levels of various phenolic pigments (summarized in Table 1; reviewed in Rozema et al., 1997; Bornman and Teramura, 1993; Tevini and Teramura, 1989). Simulations based on expected rises in UV-B radiation over the next 10–20 years indicate that plant growth, development and possibly yield will be affected. Preliminary studies indicate that these effects may also be cumulative thus affecting perennial crops more than annual crop species (Midgley et al., 1998; G. Holmes, unpublished data). Therefore, the problem is how to develop plants and crops that will perform reliably against a background of increasing levels of solar UV-B radiation.

The development of UV-B-tolerant plants requires an understanding of how UV-B is perceived by plants. UV-B radiation will be absorbed by many organic compounds within the cell, however, there is a large body of evidence indicating that responses to UV-B are downstream events of absorption by a specific UV-B photoreceptor(s) (see Ballere et al., 1991; Jansen et al., 1998; Jordan, 1996). One key question is the identification of such photoreceptors.

How absorption of UV-B radiation leads to a particular response and the signalling pathways involved are also key questions. Central to photoreceptor-mediated responses appears to be the role of reactive oxygen species (ROS) and they may form a link between non-specific effects of UV-B and specific responses such as increases in the levels of antioxidant enzymes (A.-H.-Mackerness and Jordan, 1998; A.-H.-Mackerness et al., 1998; Surplus et al., 1998).

Plant Responses to Environmental Stress, edited by M.F. Smallwood, C.M. Calvert and D.J. Bowles.
© 1999 BIOS Scientific Publishers, Oxford.

Table 1. Summary of the effects of UV-B radiation on plants

Morphological changes

 Reduction in root/shoot area, plant height, leaf length and area
 Increases in leaf thickness and axillary branching
 Changes in germination, senescence and seed production
 Changes in plant architecture and canopy structure

Effects on macromolecules

 Increases in rate of DNA damage
 Peroxidation of unsaturated fatty acids and changes in membrane lipid composition
 Decreases in carbohydrate levels

Effects on photosynthesis

 Increased turn-over of D1 and D2 proteins and decreases in efficiency of photosystem II
 Decreases in activity of Rubisco
 Decreases in chloroplast membrane integrity
 Reduction in expression of key photosynthetic genes
 Changes to chloroplast ultrastructure
 Reduction in chlorophyll and carotenoids

Protection mechanisms

 Increases in level of UV-B screening pigments (flavonoids and anthocynins)
 Increases in antioxidant enzymes and compounds
 Increases in DNA photorepair mechanisms

Rubisco, ribulose-1,5-bisphosphate carboxylase/oxygenase.

In addition, the mechanism(s) by which responses to UV-B are affected by longer wavelength radiation need to be ascertained. Higher levels of background light ameliorate the effects of UV-B radiation (A.-H.-Mackerness and Jordan, 1998; A.-H.-Mackerness *et al.*, 1998; Deckmyn *et al.*, 1994) and clearly understanding this mechanism(s) would provide valuable insight into how greater UV-B tolerance might be engineered.

With respect to bottlenecks, before identification of specific UV-B photoreceptor(s), it is difficult to prepare definitive strategies to protect plants from future rises in UV-B levels. Furthermore, at present we do not understand enough about how plants integrate responses to UV-B with responses to other environmental parameters, limiting the accuracy of projections from results obtained with plants in the laboratory to responses of plants in natural conditions. Solving these two fundamental, and very different problems would provide a foundation for engineering more tolerant crops.

2. Changes in gene expression in response to UV-B exposure

We have shown that UV-B exposure leads to dramatic changes in gene expression. Transcription and synthesis of key photosynthetic proteins are reduced after exposure to UV-B radiation. In contrast, there is an increase in the expression of genes encoding a range of proteins necessary for potential protective mechanisms. The following section provides a brief review of these changes in gene expression in response to UV-B exposure.

2.1 Photosynthetic genes

Recent studies have shown that the primary deleterious effects of UV-B on the efficiency of the photosynthetic apparatus and protein levels can be attributed to specific reductions in expression of key photosynthetic genes – these changes in gene expression are detectable within hours, but lead to reductions in protein levels and activity over a period of days (see A.-H.-Mackerness *et al.*, 1997). Transcripts for chloroplast proteins encoded on both the nuclear (*Lhcb* and *RbcS*) and chloroplast (*psbA* and *rbcL*) genomes are reduced in response to UV-B exposure. This UV-B response is dose dependent, but for a given dose of UV-B, nuclear-encoded transcripts are reduced much more severely than chloroplast-encoded transcripts. For example, in pea, the transcripts for the nuclear-encoded *Lhcb* are reduced by 80% of control levels after 4 days of UV-B exposure while the transcripts for the chloroplast-encoded *psbA* are reduced by only 10% over the same time period.

The decline in the steady-state levels of the nuclear-encoded genes such as *Lhcb* and *RbcS* mRNA correspond with the subsequent reductions in protein levels. These results, in conjunction with nuclear run-off experiments, suggest strongly that there is transcriptional control of nuclear-encoded genes by UV-B radiation. In contrast, the level of chloroplast-encoded transcripts such as *psbA* do not parallel the loss in protein levels, which occur prior to any reduction in transcript levels. UV-B exposure enhances both the rate of synthesis and degradation of the D1 polypeptide, strongly indicating that the primary level of UV-B control is post-translational (for more details see A.-H.-Mackerness *et al.*, 1997).

2.2 Defence mechanisms

It is not only photosynthetic genes that are regulated in response to UV-B stress. It is generally accepted that plants have evolved a number of protective mechanisms against UV-B-induced damage, amongst which is the up-regulation of various 'defence'-associated genes. The most important/studied protective mechanisms employed by plants in response to UV-B exposure are UV-B attenuation by UV-B-absorbing pigments, induction of antioxidant systems to minimize oxidative damage and repair of UV-B-induced DNA damage.

UV-B attenuation by UV-B absorbing pigments. More general protection is afforded by the production of protective pigments such as flavonoids and sinapic esters which limit the penetration of UV-B within the plant tissues (reviewed in Caldwell *et al.*, 1983). Studies using plants defective in flavonoid biosynthesis have illustrated that a few specific flavonoid compounds are particularly important in UV-B tolerance (Landry *et al.*, 1995) and have been shown to protect photosynthetic transcripts against UV-B radiation (Jordan *et al.*, 1998). The increase in the level of these protective pigments, in response to UV-B radiation, is due to a co-ordinated increase in the expression and activity of enzymes of the phenylpropanoid pathway at the time when photosynthetic gene expression decreases. Nuclear run-off experiments have shown that transcription rates of the phenylpropanoid pathway genes rise on response to UV-B exposure (Chappell and Hahlbrock, 1984) and UV-B responsive elements have been identified within the promoters of several of these genes (reviewed in Jenkins, 1997).

Induction of antioxidant systems. Limiting the penetration of UV-B within the tissue does not overcome the immediate stress the plant must endure. UV-B exposure leads to the generation of ROS and increases in the concentration of lipid peroxides. Connected to this oxidative stress, there are changes in stress-responsive mRNA transcripts and in the activities of antioxidant enzymes such as glutathione reductase, ascorbate peroxidase and superoxide dismutase (A.-H.-Mackerness and Jordan, 1998; Strid *et al.*, 1994). The precise defence mechanisms have only recently begun to be determined but it is clear that different species respond through the induction of different sets of antioxidants. Unfortunately, the use of transgenic plants over- or under-expressing various antioxidant enzymes have not provided useful information but have further highlighted the complexity and variation in responses between different plant species.

Repair of UV-B-induced DNA damage. DNA strongly absorbs UV-B radiation resulting in the formation of a variety of lesions, including most frequently cyclobutane pyrimidine dimers (CPDs) and pyrimidine (6–4) pyrimidone photoproducts (6–4 photoproducts) (reviewed in Britt, 1996). These are known to inhibit DNA and RNA polymerase and thus prevent transcription. To overcome this, cells contain a number of DNA repair mechanisms. With respect to UV-B-induced DNA damage, photo-reactivation is the most important and is carried out by a set of photo-activated DNA repair enzymes, the photolyases, that utilize a range of wavelengths in the UV-A and blue regions (300–500 nm). Photo-reactivation is efficient, economical and error-free and has been reported from a range of plant species (Britt, 1996) although to date, photolyase genes have only been isolated from *Arabidopsis* (Ahmad *et al.*, 1997). Expression studies have demonstrated that this *Arabidopsis* photolyase gene was not expressed in etiolated tissue but was induced by blue and UV-B light, matching photo-reactivation activity in plants.

A number of other genes associated with defence against pathogen attack and wounding are also up-regulated in response to UV-B. These include genes of both the acidic and basic type pathogenesis-related proteins and defencin (Surplus *et al.*, 1998; A.-H.-Mackerness *et al.*, unpublished data). The role of these proteins in protection against UV-B exposure is not yet clear.

3. Factors influencing effects of UV-B on gene expression

The interactions between UV-B and other stresses have not been studied in depth (Bornman and Teramura, 1993). Clearly responses to UV-B should not be considered in isolation and this is best illustrated by the influence of background light and developmental stage on the plants' response to UV-B exposure. In pea, where these differences are best characterized, older leaves become damaged by UV-B faster and to a greater extent than younger buds. Similarly, the mRNA transcripts for chloroplast proteins are maintained in young buds and are not down-regulated to the same extent as they are in older leaves (A.-H.-Mackerness and Jordan, 1998; A.-H.-Mackerness *et al.*, 1998).

In addition to the developmental stage of tissue studied, the level of background light (photosynthetically active radiation, PAR: 400–700 nm) provided throughout the UV-B treatment also has a profound influence on the sensitivity of plants to UV-B radiation and the damage on the photosynthetic apparatus. Several studies have indicated that the effectiveness of UV-B is dependent on the PAR provided prior to and during the UV-B treatment (A.-H.-Mackerness *et al.*, 1998; Bornman and Teramura, 1993; Rozema *et al.*, 1997; Tevini and Teramura, 1989). Growth under high PAR leads to morphological changes which result in increased tolerance to subsequent UV-B exposure. The mechanisms behind the effects of high PAR concomitant with UV-B exposure, after growth under low PAR (high light (HL) protection), have been more difficult to explain. Initially studies indicated that increases in photolyase activity conferred the protective effect of high PAR. However, recent studies have indicated that the activity of these enzymes is saturated at very low light levels. It seems likely that increases in photosynthesis alone, under these high PAR conditions, are sufficient to account for this protection, which is effective at the physiological, biochemical and molecular levels (A.-H.-Mackerness and Jordan, 1998; A.-H.-Mackerness *et al.*, 1998).

4. Signal transduction pathways

The signal pathways leading to changes in gene expression in response to UV-B are still largely unknown. Studies using cell cultures and various pharmaceutical compounds have provided some insight into the potential components of these pathways (Jenkins, 1997). However, the information obtained from such systems is limited and not necessarily a reflection of events within whole plants. The realization that plant responses to UV-B stress share components with responses to various biotic stresses has provided a starting point for determining the signal pathways involved in response to UV-B stress (*Figure 1*). Recent studies have indicated that, as in biotic stress, ROS are involved in UV-B signal pathways leading to the down-regulation of photosynthetic genes and

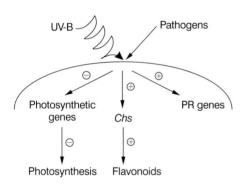

Figure 1. Schematic representation of some of the common responses to both UV-B exposure and pathogen infection.

up-regulation of the acidic-type pathogenesis-related (PR) genes (*Figure 2*; A.-H.-Mackerness *et al.*, 1998; A.-H.-Mackerness and Jordan, 1998; Surplus *et al.*, 1998). In addition, salicylic acid (SA) was shown to also increase in response to UV-B exposure, and as a result of ROS generation, leading in turn to the rise in PR transcripts (*Figure 2*; Surplus *et al.*, 1998).

The realization of the important role of ROS in regulating photosynthetic gene

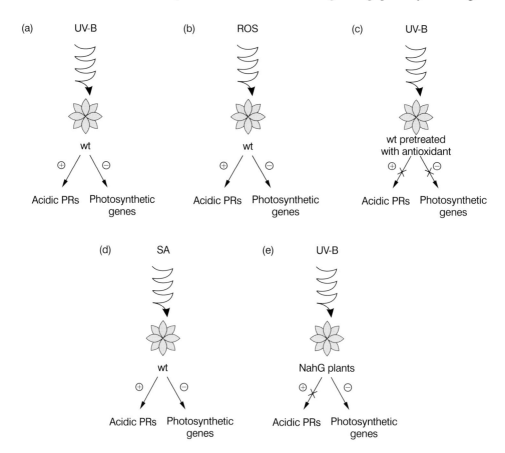

Figure 2. Schematic representation of the role of reactive oxygen species (ROS) and salicylic acid (SA) in signal pathways leading to the down-regulation of photosynthetic genes and the up-regulation of acidic-type pathogenesis-related (PR) genes in response to UV-B radiation. (a) UV treatment of wild-type (wt) *Arabidopsis* leads to an increase (+) in PR transcripts and a decrease (–) in photosynthetic transcripts. (b) Treatment of wt plants with oxidants (e.g. 3-AT, paraquat: ROS) mimics the effects of UV-B. (c) Treatment of wt plants with antioxidants (e.g. ascorbic acid or PDTC), prior to UV-B treatment, prevents both the increase in PR transcripts and the decrease in photosynthetic transcripts. Thus (b) and (c) strongly indicate that ROS play a role in regulation of both sets of genes in response to UV-B radiation. (d) Treatment of wt plants with SA mimics the effects of UV-B. (e) UV-B treatment of transgenic NahG plants, which cannot accumulate SA, does not lead to an increase in PR transcripts but the decrease in photosynthetic transcripts still occurs. Thus, (d) and (e) indicate that although SA can lead to the down-regulation of photosynthetic genes, SA is only involved in the up-regulation of PR genes in response to UV-B radiation.

expression, in response to UV-B, led to the proposal that there might be a common mechanism which could explain developmental sensitivity and HL protection against UV-B damage (A.-H.-Mackerness *et al.*, 1998). The levels of ROS within a tissue are a balance between ROS generation and removal and thus the extent of UV-B damage is dependent on the antioxidant capacity of the tissue studied. The antioxidant capacity of a cell is primarily dependent on the rate of photosynthesis which is in turn dependent, amongst other parameters, on the developmental stage of the tissue and the level of background irradiance. Thus younger tissues and plants under high PAR have higher antioxidant capacities and are thus better able to remove ROS and are consequently less sensitive to UV-B-induced stress (for further details see A.-H.-Mackerness *et al.*, 1998). The role of ROS in UV-B signalling is now established in several species, however, the origin of these ROS in response to UV-B exposure has remained elusive.

Clearly, although we are beginning to understand more about how plants respond to UV-B radiation, we are still far from appreciating exactly how increases in UV-B levels will affect plants, how plants respond to UV-B and, importantly, how we can combat any of the detrimental effects of increasing UV-B exposure. Only when we understand these processes more completely will we be able to devise satisfactory strategies to protect our crops, trees and vegetation in a world where stratospheric ozone continues to decrease and UV-B levels will continue to rise.

Acknowledgements

The research is supported by the BBSRC. We are grateful to Margaret Jones for preparation of the figures and to Drs Richard Napier and Vicky Buchanan-Wollaston for critical reading of this manuscript and many helpful comments.

References

Ahmad, M., Jarillo, J.A., Klimczak, L.J., Landry, L.G., Peng, T., Last, R.L. and Cashmore, A.R. (1997) An enzyme similar to animal type II photolyase mediates photoreactivation in *Arabidopsis*. *Plant Cell* **9**: 199–207.

A.-H.-Mackerness, S. and Jordan, B.R. (1999) Changes in gene expression in response to UV-B induced stress. In: *Handbook of Plant and Crop Stress* (ed. M. Pessarakli). Marcel Dekker Inc, New York, pp. 749–768.

A.-H.-Mackerness, S., Jordan, B.R. and Thomas, B. (1997) UV-B effects on the expression of genes encoding proteins involved in photosynthesis. In: *Plants and UV-B: Responses to Environmental Change* (ed. P.J. Lumsden). Cambridge University Press, Cambridge, pp. 113–134.

A.-H.-Mackerness, S., Surplus, S.L., Jordan, B.R. and Thomas, B. (1998) Effects of supplementary UV-B radiation on photosynthetic transcripts at different stages of development and light levels in pea: role of ROS and antioxidant enzymes. *Photochem. Photobiol.* **68**: 88–96.

Ballere, C.L., Barnes, P.W. and Kendrick, R.E. (1991) Photomorphogenic effects of UV-B radiation on hypocotyl elongation in wild type and stable phytochrome-deficient mutant seedlings of cucumber. *Physiol. Plant.* **83**: 652–658.

Bornman, J.F. and Teramura, A.H. (1993) Effects of UV-B radiation on terrestrial vegetation. In: *Environmental UV Photobiology* (eds A.R. Young, L.O. Bjorn and J. Moan). Plenum Press, London, pp. 427–472.

Britt, A.B. (1996) DNA damage and repair in plants. *Annu. Rev. Plant Physiol. Plant Mol. Biol.* **47:** 75–100.

Caldwell, M.M., Robberecht, R. and Flint, S.D. (1983) Internal filters: prospect for UV-acclimation in higher plants. *Physiol. Plant.* **58:** 445–450.

Chappell, J. and Hahlbrock, K. (1984) Transcription in plant defence genes in response to UV light or fungal elicitor. *Nature* **311:** 76–78.

Deckmyn, G., Martens, C. and Impens, I. (1994) The importance of the ratio of UV-B/PAR during leaf development as determining factor of plant sensitivity to increased UV-B irradiance: effects on growth, gas exchange and pigmentation of bean plants. *Plant Cell Environ.* **17:** 295–301.

Jansen, M.A.K., Baba, V. and Greenberg, B. (1998) Higher plants and UV-B radiation: balancing damage, repair and acclimation. *Trends Plant Sci.* **3:** 131–135.

Jenkins, G.I. (1997) UV and blue light signal transduction in *Arabidopsis. Plant Cell Environ.* **20:** 773–778.

Jordan, B.R. (1996) The effects of UV-B radiation on plants: a molecular prospective. In: *Advances in Botanical Research* (ed. J.A. Callow). Academic Press, New York, pp. 97–162.

Jordan, B.R., James, P.E. and A.-H.-Mackerness, S. (1998) Factors affecting UV-B induced changes in *Arabidopsis* gene expression: the role of development, protective pigments and the chloroplast signal. *Plant Cell Physiol.* **39:** 769–778.

Kerr, J.B. and McElroy, C.T. (1993) Evidence for large upward trends of UV-B radiation linked to ozone depletion. *Science* **262:** 1032–1034.

Landry, L.G. *et al.* (1995) *Arabidopsis* mutants lacking phenolic sunscreens exhibit enhanced UV-B injury and oxidative damage. *Plant Physiol.* **109:** 1159–1166.

Madronich, S. *et al.* (1995) Changes in UV radiation reaching the earth's surface. *AMBIO* **24:** 143–152.

Midgley, G.F., Wand, J.E. and Musil, C.F. (1998) Repeated exposure to enhanced UV-B radiation in successive generations increases developmental instability in desert annual. *Plant Cell Environ.* **21:** 437–442.

Rozema, J., van de Staaij, J., Bjorn, L.O. and Caldwell, M. (1997) UV-B, an environmental factor in plant life: stress and regulation. *TREE* **12:** 22–28.

Strid, A., Chow, W.S. and Anderson, J.M. (1994) UV-B damage and protection at the molecular level in plants. *Photosynth. Res.* **39:** 475–489.

Surplus, S.L., Jordan, B.R., Murphy, A.M., Carr, J.P., Thomas, B. and A.-H.-Mackerness, S. (1998) UV-B induced responses in *Arabidopsis thaliana*: role of salicylic acid and ROS in the regulation of transcripts and acidic PR proteins. *Plant Cell Environ.* **21:** 685–694.

Tevini, M. and Teramura, A.H. (1989) UV-B effects on terrestrial plants. *Photochem. Photobiol.* **50:** 479–487.

Chapter 4

The role of hydrogen peroxide and antioxidants in systemic acclimation to photo-oxidative stress in *Arabidopsis*

Stanislaw Karpinski, Helen Reynolds, Barbara Karpinska, Gunnar Wingsle, Gary Creissen and Philip Mullineaux

1. Background

Land plants are sessile in nature and have developed sophisticated mechanisms which allow for immediate, acclimatory and adaptive responses to changing environments. The prevailing light conditions a plant encounters determines the efficiency of its photosynthetic apparatus such that light capture, energy transduction and assimilation are optimized. However, rapid changes in light intensity always occur in the natural environment and may cause an imbalance such that the light energy absorbed through photochemistry is in excess of that which is able to be consumed by metabolism, principally CO_2 fixation. Other environmental factors such as the prevailing temperature, nutritional and water status of the plant have a substantial impact on setting the level of photochemical energy which is required by metabolism (Baker, 1994; Foyer, 1996; Huner *et al.*, 1998; Krause, 1994).

In higher plants, dissipation of excess photochemical energy (EPE) is an immediate and finely tuned response which occurs through heat irradiation, alternative sinks for photosynthetic electrons and ultimately down-regulation of photosystem II (PSII) (Baker, 1994; Foyer, 1996; Huner *et al.*, 1998; Krause, 1994). The photoreduction of oxygen is an important alternative sink for the consumption of excess energy but is associated with an increase in the generation of reactive oxygen intermediates (ROIs) such as hydrogen peroxide (H_2O_2), the superoxide anion ($O_2{\cdot}^-$), the hydroxyl radical ($\cdot$OH) and singlet oxygen ($^1O_2$) (Foyer, 1996). If accumulation of ROIs under conditions of EPE exceeds the capacity of enzymic and non-enzymic antioxidant systems to remove them, then oxidative damage to the photosynthetic apparatus ensues, which leads to cell death. This is manifested at the whole plant level by the appearance of chlorotic lesions on damaged leaves (*Figure 1b*). However, ROIs may also play a positive role in the response to EPE by initiating an increase in the rate of degradation of the D1 protein, a key component of PSII reaction centres (Krause, 1994). This causes

Plant Responses to Environmental Stress, edited by M.F. Smallwood, C.M. Calvert and D.J. Bowles.
© 1999 BIOS Scientific Publishers, Oxford.

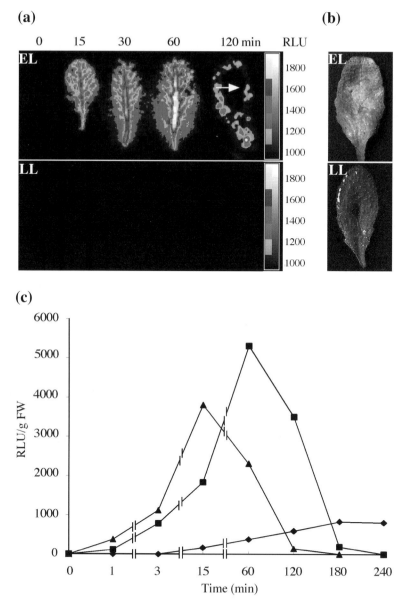

Figure 1. Excess light-induced photooxidative damage and induction of *APX2-LUC* and *APX2* in transgenic leaf tissue of *Arabidopsis* (a, b and c). (a) CCD camera images of relative LUC activity in detached leaves exposed to different time EL (arrow indicates the dead zone of the leaf).
(b) Appearances of chlorotic lesions on detached leaves after 2 h excess light (EL; 2700 ± 300 μmol of photons $m^{-2} s^{-1}$). Leaves of *APX2-LUC* transgenic plants grown in LL (200 ± 30 μmol of photons $m^{-2} s^{-1}$; control) were exposed to EL blast for up to 2 h. (c) Induction kinetic of *APX2-LUC* in leaves exposed to different light intensity; 1250 (-◆-), 2500 (-■-) and 3500 (-▲-) μmol of photons $m^{-2} s^{-1}$.

photoinhibition of photosynthetic electron transport which may be a protective mechanism in such conditions. The possibility of both positive and negative roles for ROIs in EPE suggests that it is more appropriate to view the equilibrium between processes which produce ROIs and antioxidant processes which destroy them, rather than the levels of these antagonists *per se* (Foyer, 1996).

The immediate responses to EPE may lead to a whole plant acclimation which could include an alteration to the photosynthetic capacity of new leaves (Huner *et al.*, 1998) in which ROIs also could play a role (Foyer *et al.*, 1997). Therefore in the natural environment, how well plants tolerate a wide spectrum of abiotic stresses may be determined by their ability to deal with EPE before oxidative damage to cellular structures becomes a significant problem.

2. The problem and scientific approach

The above considerations imply that some form of intercellular communication must occur between stressed leaves, unstressed leaves and those still undergoing development, although to our knowledge no such systemic mechanism for responses to EPE have been described. Using the experimental system described below, we set out to address this problem and to make a linkage between immediate and acclimatory responses to EPE.

Experimentally, excess light (EL) applied to low-light (LL) adapted *Arabidopsis* plants for up to 1 h causes rapid EPE and an increase in ROI production leading to reversible photoinhibition (Karpinski *et al.*, 1997; Russell *et al.*, 1995). Surprisingly, it was found that such a chloroplast-localized oxidative stress only induced genes encoding key components of the cytosolic ROI scavenging system. One of these genes, *APX2*, encoding an ascorbate peroxidase (Santos *et al.*, 1996; EC 1.11.1.11) isoform, is unique among oxidative stress-responsive genes in being induced only under EL, with no detectable expression in LL (Karpinski *et al.*, 1997). The induction of this nuclear gene is regulated by changes in the activity of PSII, probably by rapid changes in the redox status of the plastoquinone (PQ) pool. Moreover, the major cellular antioxidant glutathione (Noctor *et al.*, 1998) blocked the induction of this gene by EL, implying that redox changes in the cellular glutathione pool(s) may play a role in chloroplast-to-nucleus communication (Karpinski *et al.*, 1997).

Leaves from *Arabidopsis* plants transformed with an *APX2* promoter-luciferase gene (*LUC*) fusion had no detectable luciferase expression when grown under LL conditions but on challenging with EL for a minimum of 15 min induced luciferase activity which could be imaged using a CCD camera. The induction of the *APX2-LUC* transgene mirrored the induction of the native *APX2* gene in the same plants, as previously described (Karpinski *et al.*, 1997). Using a more sensitive *in vitro* assay for luciferase activity, induction of *APX2-LUC* could be detected after only 3 min exposure to EL at 2500 μmol m^{-2} s^{-1} (*Figure 1*). The induction of luciferase activity in the *APX2-LUC* transgenic plants could be taken as a convenient measure of the activation of *APX2* expression. After 2 h exposure to EL, the leaves suffered irreversible photoinhibition and this was associated with a loss of *APX2-LUC* expression (*Figure 1*) over most of the leaf area and no detectable *APX2* mRNA in the northern blots. After 24 h under LL conditions, such leaves developed chlorosis.

The induction of *APX2-LUC* by EL was abolished if leaves were first infiltrated with glutathione (γ-L-glutamylcysteinylglycine; Noctor *et al.*, 1998), either in the reduced form (GSH) or the oxidized form (GSSG). The same effect could be achieved by pre-infiltrating leaves with ascorbate and the signal could be diminished by infiltrating leaves with catalase. EL-induction of luciferase activity was not inhibited when superoxide dismutase was pre-infiltrated into the leaves (*Figure 2*). Taken together these data suggest that H_2O_2, but not $O_2^{\cdot-}$, was implicated in the EL-induced expression of *APX2*. It is possible that glutathione and ascorbate could cause specific effects in the inhibition of *APX2* expression and therefore both be components of a regulatory system. However, a simpler explanation is that both compounds behaved in a common way as antioxidants and exerted their inhibition of *APX2* expression by removing its inducer, H_2O_2. It has been shown that GSSG can be rapidly taken up into cells, probably by a specific transporter at the plasmalemma (Jamai *et al.*, 1996), and rapidly reduced to GSH (Wingsle and Karpinski, 1996). Therefore, the effect of GSSG was expected to be similar to that observed for GSH. H_2O_2 on its own did not induce the expression of *APX2-LUC* to the same level as EL such that luciferase activity could be imaged, but could be detected by the more sensitive *in vitro* assay (*Table 1*).

It has previously been demonstrated that the induction of *APX2* transcript levels is regulated by redox events in the proximity of PSII, probably the redox status of PQ (Karpinski *et al.*, 1997). This conclusion was arrived at from the treatment of detached leaves with the photosynthesis electron transport inhibitors, 2,5-dibromo-3-methyl-6-isopropyl-*p*-benzoquinone (DBMIB) and 3-(3,4-dichlorophenyl)-1,1-dimethylurea (DCMU) before exposure to EL. DCMU and DBMIB block the reduction and oxidation of PQ, respectively, and these effects are indicated by the chlorophyll *a* fluorescence parameters, F_v/F_m and q_p (*Table 1*). The ratio F_v/F_m is the dark-adapted chlorophyll *a* fluorescence parameter which indicates the maximum photochemical efficiency of PSII

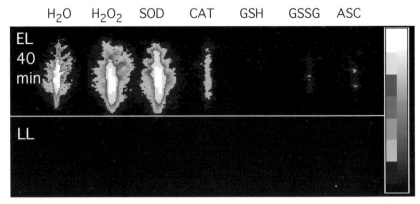

Figure 2. Regulation of *APX2-LUC* expression in transgenic *Arabidopsis* leaf tissue. CCD camera image of luciferase activity (RLU) in detached leaves treated with water (control), 10 mM H_2O_2, superoxide dismutase (200 units of SOD), catalase (200 units of CAT), reduced and oxidized glutathione (10 mM GSH and GSSG) and ascorbate (25 mM ASA) for up to 2 h in LL and then exposed to EL for 40 min.

Table 1. Photosynthetic parameters and luciferase activity in *APX2-LUC* transgenic plants exposed to EL

| Treatment | Low light (2 h) | | | Low light + excess light (2 h + 1 h) | | |
	F_v/F_m	q_p	Luciferase activity	F_v/F_m	q_p	Luciferase activity
H$_2$O$_2$	0.89 ± 0.02	0.9 ± 0.03	306 ± 45	0.6 ± 0.05	0.74 ± 0.03	6413 ± 534
DBMIB	0.81 ± 0.04	0.84 ± 0.05	388 ± 58	0.49 ± 0.04	0.62 ± 0.04	5347 ± 457
DCMU	0.41 ± 0.08	0.11 ± 0.07	27 ± 18	0.29 ± 0.05	0.03 ± 0.03	57 ± 49
DCMU+H$_2$O$_2$	0.47 ± 0.07	0.28 ± 0.03	84 ± 25	0.31 ± 0.04	0.08 ± 0.03	56 ± 31
H$_2$O	0.83 ± 0.02	0.85 ± 0.02	10 ± 10	0.42 ± 0.02	0.5 ± 0.01	5054 ± 623

Detached leaves were treated with hydrogen peroxide (H$_2$O$_2$), 2,5-dibromo-3-methyl-6-isopropyl-*p*-benzoquinone (DBMIB) or 3-(3,4-dichlorophenyl)-1,1-dimethylurea (DCMU) and exposed to different light conditions. LL, low light (200 µmol of photons m^{-2} s^{-1}); EL, excess light (2800 µmol of photons m^{-2} s^{-1}). Luciferase activity is in RLU g^{-1} fresh weight. F_v/F_m, q_p and *APX2-LUC* values were determined in leaves treated with 10 mM H$_2$O$_2$, 10 µM DCMU, 15 µM DBMIB or water. Parameters were measured in three different leaves obtained from three independent experiments ($n = 9 ± SD$) with duplicate readings taken for each data point.

reaction centres and q_p reflects the extent of reduction of the primary electron acceptor quinone, Q_A, which is associated with the PSII complex (Krause, 1988). The *APX2-LUC* transgene was induced slightly in LL by DBMIB and DCMU blocked the induction of luciferase activity in EL, confirming previous observations with *APX2* mRNA levels (Karpinski *et al.*, 1997). H$_2$O$_2$ did not relieve the block by DCMU on EL-mediated induction of *APX2-LUC* and furthermore, the mild inductive effect of H$_2$O$_2$ observed in LL was blocked by the presence of DCMU (*Table 1*). These data indicated that even in LL the slight induction by H$_2$O$_2$ of *APX2-LUC* still required a functional PSII.

Treatment of leaves with H$_2$O$_2$ and then exposure to EL caused a significantly greater induction of *APX2-LUC* activity than control EL exposures alone (*Table 1*). The use of the fluorescence parameters to monitor the response of the leaves to the inhibitor and H$_2$O$_2$ treatments also revealed that H$_2$O$_2$ alone consistently brought about a lesser degree of photoinhibition (*Table 1*). This surprising observation was investigated in more detail in a series of time-course experiments which revealed that detached leaves pre-treated with H$_2$O$_2$ showed a slower decline in maximal PSII efficiency under EL conditions than water-treated control leaves, indicating that the pro-oxidant status of the leaf may be a crucial factor in adjusting to EPE. Protective effects by H$_2$O$_2$ have been described for maize seedlings chilled in the dark (Prasad *et al.*, 1994) and thermotolerance of potato nodal explants in tissue culture (Delgado-Lopez *et al.*, 1998), and have been explained by the triggering of various stress defence mechanisms (Foyer *et al.*, 1997). Our data do not indicate whether H$_2$O$_2$ has a direct or indirect effect but do strongly point to a role of this compound in the acclimation to conditions which invoke EPE. The opposite effects of exacerbated PSII inhibition upon treatment with the antioxidant glutathione (*Figure 2*; Karpinski *et al.*, 1997), which reduces H$_2$O$_2$ by enzymic and non-enzymic reactions (Mullineaux and Creissen, 1997), are consistent with such a role for H$_2$O$_2$.

We considered it unlikely that in our procedures, catalase could readily penetrate into cells on a scale which would allow us to observe a loss of luciferase signal at the whole leaf level. Therefore, the catalase-mediated inhibition of *APX2-LUC* induction strongly suggested that H_2O_2 involved in the response to EPE, at least in part, was extracellular and could act as a cell-to-cell messenger, amplifying the *APX2* response. To investigate this further, detached leaves were subjected to EL for 40 min in the top quarter of the leaf only and luciferase activity was imaged 20 min later. While luciferase activity was detected in the adjacent shaded areas no activity could be imaged in the rest of the leaf, save for a localized region at the cut surface of the petiole. However, in sections of leaf 1 cm away from the LL/EL boundary, luciferase activity could be detected in the LL zone by the more sensitive *in vitro* assay (443 ± 155 RLU g^{-1} fresh wt; $n = 9 \pm$ SD). This was accompanied by a reduction in maximal PSII efficiency ($F_v/F_m = 0.75 \pm 0.03$ compared with LL control $F_v/F_m = 0.84 \pm 0.04$; $n = 9 \pm$ SD). Deliberate wounding of leaves in the lowest quarter and exposure of the top quarter to EL increased the area of leaf displaying luciferase activity, which was also more intense than in non-wounded detached leaves. The more severe wounding alone in these experiments also failed to induce *APX2-LUC*.

When mature leaves in sections of whole rosettes, covering approximately one-third of the total plant area, were exposed to EL for 30 min, luciferase activity was readily detected in the apical region and lower sections of young leaves. The activation of *APX2-LUC* in whole plants in regions remote from exposure to EL suggested that systemic signalling occurred and was most active in immature growing leaves. This notion is supported by the experiments with detached and wounded leaves since cells in wounded regions are stimulated to divide. That systemic signalling occurred between leaves suffering EPE and young leaves leads us to speculate that one of the immediate reactions is to initiate an adaptive response to this stress. In this sense, the *APX2-LUC* can be regarded as a visual marker gene for these events.

The systemic signalling which triggered *APX2* expression raised the question of whether it regulates acclimation of the plant to EPE. To address this question, leaves on one side of rosettes were exposed to EL (hereafter called 1° leaves), but leaves from the LL side of the rosette (hereafter called 2° leaves) were measured for their H_2O_2 content, luciferase activity, chlorophyll *a* fluorescence parameters and abundance of *APX2* mRNA (*Table 2*). These data show that the 2° leaves induced *APX2* expression to approximately 15% of the levels in EL-exposed 1° leaves, as measured by both luciferase activity and the relative abundance of *APX2* mRNA. This was similar to the luciferase activity induced in H_2O_2-treated detached leaves (*Table 1*). This partial activation of *APX2* expression was associated with an increase in H_2O_2 content and no significant change in PSII efficiency (*Table 2*). Under these conditions, the 1° leaf showed clear signs of photo-oxidative stress (*Table 2*). Subsequently, a full exposure of partial EL-treated rosettes to further EL for 30 min, exacerbated the photo-oxidative stress in 1° leaves but 2° leaves showed increased tolerance. They displayed only a slight reduction in PSII efficiency, no further increase in H_2O_2 levels and a level of *APX2* induction greater than in 1° leaves (*Table 2*). Taken together, these data indicate that a systemic signal can regulate an acclimatory response to EPE-inducing environmental conditions.

Table 2. Systemic induction of *APX2* is associated with increased tolerance to excess light (EL) in secondary (2°) leaves

Treatment	F_v/F_m	q_p	H_2O_2 content (μmol g^{-1} fresh wt)	Luciferase activity (RLU g^{-1} fresh wt)	APX2 RNA level
LL leaf	0.83 ± 0.02	0.85 ± 0.02	4.2 ± 0.8	<10	−
EL 1° leaf	0.62 ± 0.04	0.72 ± 0.03	6.8 ± 0.7	5689 ± 386	++++
LL 2° leaf	0.79 ± 0.03	0.81 ± 0.03	5.7 ± 0.5	780 ± 238	+
EL no. 2 to EL 1° leaf	0.4 ± 0.02	0.5 ± 0.04	7.3 ± 0.7	6282 ± 421	++++
EL no. 2 to LL 2° leaf	0.76 ± 0.03	0.79 ± 0.03	5.9 ± 0.6	6747 ± 589	++++

Rosettes of low-light (LL; 200 μmol m^{-2} s^{-1}) short-day grown *Arabidopsis* were exposed at a tangent covering one-quarter of its area with excess light (EL; 2500 μmol m^{-2} s^{-1}) for 30 min. Primary (1°) leaves exposed to the light stress and secondary (2°) leaves from the same rosette which had not been exposed to EL were measured for the above parameters. Different partial EL-treated rosettes were then exposed to a further 30 min of EL (EL no. 2) over the whole area of the rosette and equivalent 1° and 2° leaves were assayed. These data are combined from three independent experiments with three leaves per treatment ($n = 9 \pm$ SD) with duplicate readings taken for each data point.

3. Conclusion

The work presented here allows a unified view of the response to fluctuating environmental conditions which elicit EPE; bringing together immediate and acclimatory responses. When a leaf experiences a set of conditions which promote EPE, such as EL conditions, the immediate response is an intracellular signalling of antioxidant defence genes elicited by redox changes in the proximity of PSII, but still followed by a subsequent increase in H_2O_2 levels and elaboration of photoinhibition. Prolonged exposure to such stress leads to the death of such leaves. However, leaves suffering EPE also produce a systemic signal, a component of which is H_2O_2, which sets up an acclimatory response in unstressed regions of the plant. This signalling leads to an increased capacity to tolerate further episodes of EPE-induced photo-oxidative stress by a remote activation of antioxidant defences. Furthermore, taking into account that redox events in PSII are a prerequisite for induction of *APX2* in LL and EL, it is concluded that information regarding PSII activity in stressed chloroplasts can be transmitted systemically to modulate PSII function in non-stressed chloroplasts of remote cells which have never experienced conditions causing EPE. We have termed this phenomenon systemic acquired acclimation.

References

Baker, N.R. (1994) Chilling stress and photosynthesis. In: *Causes of Photooxidative Stress and Amelioration of Defense Systems in Plants* (eds. C.H. Foyer and P.M. Mullineaux). CRC Press, Boca Raton, FL, pp. 127–154.

Delgado-Lopez, H., Dat, J.F., Foyer, C.H. and Scott, I.M. (1998) Induction of thermotolerance in potato microplants by acetylsalicylic acid and H_2O_2. *J. Exp. Bot.* **49**: 713–720.

Foyer, C.H. (1996) Oxygen metabolism and electron transport in photosynthesis. In: *Oxidative Stress and the Molecular Biology of Antioxidant Defenses* (ed. J. Scandalios). Cold Spring Harbor Laboratory Press, New York, pp. 587–621.

Foyer, C.H., Lopez-Delgado, H., Dat, J.F. and Scott, I.M. (1997) Hydrogen peroxide- and glutathione-associated mechanisms of acclimatory stress tolerance and signalling. *Physiol. Plant.* **100**: 241–254.

Huner, N.P.A., Öquist, G. and Sarhan, F. (1998) Energy balance and acclimation to light and cold. *Trends Plant Sci.* **3**: 224–230.

Jamai, A., Tommasini, R., Martinoia, E. and Delrot, S. (1996) Characterisation of glutathione uptake in broad bean leaf protoplasts. *Plant Physiol.* **111**: 1145–1152.

Karpinski, S., Escobar, C., Karpinska, B., Creissen, G. and Mullineaux, P.M. (1997) Photosynthetic electron transport regulates the expression of cytosolic ascorbate peroxidase genes in *Arabidopsis* during excess light stress. *Plant Cell* **9**: 627–640.

Krause, G.H. (1988) Photoinhibition of photosynthesis. An evaluation of damaging and protective mechanisms. *Physiol. Plant.* **74**: 566–574.

Krause, G.H. (1994) The role of oxygen in photoinhibition and photosynthesis. In: *Causes of Photooxidative Stress and Amelioration of Defense Systems in Plants* (eds. C.H. Foyer and P.M. Mullineaux). CRC Press, Boca Raton, FL, pp. 43–76.

Mullineaux, P.M. and Creissen, G. (1997) Glutathione reductase: regulation and role in oxidative stress. In: *Oxidative Stress and the Molecular Biology of Antioxidant Defenses* (ed. J. Scandalios). Cold Spring Harbor Laboratory Press, New York, pp. 667–713.

Noctor, G., Arisi, A.-C.M., Joanin, L., Kunert, K.J., Rennenberg, H. and Foyer, C.H. (1998) Glutathione: biosynthesis, metabolism and relationship to stress tolerance explored in transformed plants. *J. Exp. Bot.* **49**: 623–647.

Prasad, T.K., Anderson, M.D., Martin, B.A. and Stewart, C.R. (1994) Evidence for chilling-induced oxidative stress in maize seedlings and regulatory role of hydrogen peroxide. *Plant Cell* **6**: 65–74.

Russell, A.W., Critchley, C., Robinson, S.A., Franklin, L.A., Seaton, G.G.R., Chow, W.-S., Anderson, J. and Osmond, C.B. (1995) Photosystem II regulation and dynamics of the chloroplast D1 protein in *Arabidopsis* leaves during photosynthesis and photoinhibition. *Plant Physiol.* **107**: 943–952.

Santos, M., Gousseau, H., Lister, C., Foyer, C., Creissen, G., and Mullineaux, P. (1996) Cytosolic ascorbate peroxidase from *Arabidopsis thaliana* L. is encoded by a small multigene family. *Planta* **198**: 64–69.

Wingsle, G. and Karpinski, S. (1996) Differential redox regulation by glutathione of glutathione reductase and CuZn-superoxide dismutase gene expression in *Pinus sylvestris* L. needles. *Planta* **198**: 151–157.

Chapter 5

Ozone-induced genes: mechanisms and biotechnological applications

Dieter Ernst, Bernhard Grimmig, Bernd Heidenreich, Roland Schubert and Heinrich Sandermann Jr

1. Background

As a result of anthropogenic activities, concentrations of tropospheric ozone have increased during the last decades, and in summer ambient ozone concentrations are in the range between 20 and 80 nl l^{-1}, with peak episodes up to 200 nl l^{-1} in Europe and North America (Krupa *et al.*, 1995). Ozone is known to be phytotoxic, thus affecting plant health (Krupa *et al.*, 1995; Sandermann, 1996). Once ozone enters the intercellular leaf space it is converted into reactive oxygen species (ROS), and additionally triggers an oxidative burst (Schraudner *et al.*, 1998). This provokes plant reactions that are known to occur during the hypersensitive response (HR) in an incompatible plant–pathogen interaction. More generally, ozone acts as an abiotic elicitor of plant defence reactions that are known to be involved in HR and systemic acquired resistance (SAR) (Kangasjärvi *et al.*, 1994; Sandermann *et al.*, 1998). The induced stress reactions led to the model of a changed plant predisposition by ozone, resulting in an enhanced susceptibility or tolerance for a second stressor (Langebartels *et al.*, 1997; Sandermann, 1996). Gene transcripts increased by ambient ozone concentrations have first been reported for pathogenesis-related (PR) proteins (Ernst *et al.*, 1992), for proteins of the phenylpropanoid metabolism (Eckey-Kaltenbach *et al.*, 1994), and for proteins of the antioxidative system (Sharma and Davis, 1997; Willekens *et al.*, 1994). Certain photosynthetic transcripts are suppressed by ozone (Pell *et al.*, 1997). Up to now more than 40 ozone-affected transcripts have been published and reviewed (Sandermann, 1996; Sandermann *et al.*, 1998). The expression of these 'ozone-induced' genes shows a temporal and spatial hierarchy. Some genes undergo a rapid activation, whereas others are more slowly activated, even after 8–48 h (Eckey-Kaltenbach *et al.*, 1994; Sharma and Davis, 1997). In addition, localized and tissue-specific transcript accumulation have been shown (Ernst *et al.*, 1996). Ethylene and salicylic acid (SA) have been implicated in signalling ozone-induced gene expression (Sandermann, 1998; Sharma *et al.*, 1996; Tuomainen *et al.*, 1997). There is also one report on a role of jasmonic acid (JA) (Örvar *et al.*, 1997).

Plant Responses to Environmental Stress, edited by M.F. Smallwood, C.M. Calvert and D.J. Bowles.
© 1999 BIOS Scientific Publishers, Oxford.

The first ozone-responsive promoter region has been described for the resveratrol synthase promoter of grapevine (*Vst1*) (Schubert *et al.*, 1997). The purpose of this contribution is to present the current stage of ozone-induced gene regulation and some consequences for a biotechnological application.

2. Ozone-responsive promoters

Gene transcription is modulated by *trans*-acting protein factors, binding to the respective *cis*-acting elements, present in the promoter sequence. Analysis of distinct promoter fragments fused to a reporter gene allows the identification of single or multiple promoter sequence elements that affect the level of expression. The *Vst1* promoter of grapevine, triggering the expression of stilbene synthase (STS) and the *LE-ACO1* promoter of tomato, regulating 1-aminocyclopropane-1-carboxylic acid oxidase expression, are induced by ozone (Betz, 1998; Schubert *et al.*, 1997). Analysis of transgenic tobacco plants, harbouring 5′-deletion constructs of the *Vst1* promoter fused to the bacterial β-glucuronidase reporter gene, indicated that the DNA region of −430 to −280 bp is regulating ozone-induced *STS* gene expression (*Figure 1*) (Schubert *et al.*, 1997).

3. Putative regulatory sequences of the ozone-responsive *Vst1* promoter region

3.1 Myb-like recognition elements

A search for potential protein binding sites with MatInspector V2.2 (Quandt *et al.*, 1995) within the ozone-responsive region (−430 to −280 bp) revealed numerous 4 bp binding sequences, including one consensus sequence for c-Myb-binding (cattacgaagGTTGgtag, −369) (*Figure 2*). Myb-binding elements, well known in proto-oncogene families of mammals (Graf, 1992), have been found also in drought- and abscisic acid-responsive promoters (Abe *et al.*, 1997), as well as in defined boxes of phenylalanine ammonia-lyase, 4-coumaroyl-CoA ligase and chalcone synthase genes (Douglas, 1996). Transcripts of the latter genes have also been shown to be induced by ozone (Eckey-Kaltenbach *et al.*, 1994). The promoter of the tobacco *PR1-a* gene also contains a Myb-binding sequence and interacts with a recombinant Myb1 protein (Yang and Klessig, 1996). Tobacco mosaic virus (TMV) induced *PR1*, as well as *Myb1* gene expression during HR and concomitant SAR, and it has been shown that SA is involved in this regulation of gene expression (Yang and Klessig, 1996). Ozone fumigation of tobacco plants resulted in increased levels of PR protein transcripts (Ernst *et al.*,

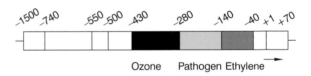

Figure 1. Schematic representation of the *Vst1* promoter of the resveratrol synthase gene. Ozone-, pathogen- and ethylene-responsive regions are indicated.

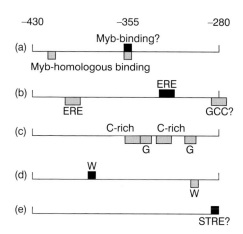

Figure 2. Schematic representation of putative *cis*-acting elements present in the ozone-responsive *Vst1* promoter of the resveratrol synthase gene. (a) Myb-like binding elements; (b) ethylene-responsive (GCC) and ethylene-responsive enhancer (ERE) elements; (c) G-boxes and C-rich domains; (d) elicitor-responsive elements (W-boxes); (e) stress-responsive element (STRE). Black boxes correspond to matrix position in sense strand, and dotted boxes to antisense orientation, respectively. All sequences are given in 5′–3′ direction. The basic sequence is taken from Schubert *et al.* (1997). Part (b) from Grimmig B. *et al.* (1997) Ozone- and ethylene-induced regulation of a grapevine resveratrol synthase promoter in transgenic tobacco. *Acta Physiologiae Plantarum*, vol. 19, pp. 467–474. Reprinted by permission of *Acta Physiologiae Plantarum*.

1992). In addition, two Myb-homologous binding sequences of the maize activator P gene (CC$^{T}/_{A}$ACC), important for flavonoid biosynthesis (Grotewold *et al.*, 1994), are also present in an inverse orientation in the ozone-responsive region (nCAACC, –414; CCAACC, –356) (*Figure 2*). Therefore it might be possible that Myb-like recognition sequences are involved in ozone-induced gene regulation as has been shown for genes responsive to pathogens (Rushton and Somssich, 1998).

3.2 Ethylene-responsive elements

Ethylene synthesis is induced during developmental plant stages, as well as by exoge-nous factors like pathogens, elicitors or wounding. As a gaseous substance ethylene can diffuse to target sites and control gene expression through a GCC-box (GAGCCGCC) *cis*-acting element (Fluhr, 1998). GCC-boxes have been found in many promoter regions of several PR genes (Eyal *et al.*, 1993) important for plant protection against pathogens. Exposure of plants to ozone results in a rapid ethylene emission (Kangasjärvi *et al.*, 1994; Sandermann, 1996; 1998) and it has been shown that ozone causes a selective activation of genes involved in ethylene biosynthesis (Tuomainen *et al.*, 1997). In addition, the full-length *Vst1* promoter is also induced by ethylene (Grimmig *et al.*, 1997). However, in its ozone-responsive region no GCC-box was found. Interestingly, an inverse GCC-box-like element (GAGCCcCt) is present at posi-tion –284 to –277 (*Figure 2*). This element has been disrupted in the ozone-insensitive

–280 bp 5'-deletion transformant (Schubert *et al.*, 1997). For TMV-induced expression of tobacco β-*glucanase* promoter activity a single inverted GCC-box is necessary, and base substitution at the 3'-CCG to 3'-TAT abolished virus-induced promoter activity (Livne *et al.*, 1997). Therefore it is not substantiated whether the 3'-tCc end of the GCC-box-like motif contributes to ethylene signalling. On the other hand, two repeated ethylene-responsive enhancer-like (ERE) elements are present in the ozone-responsive *Vst1* promoter region (Grimmig *et al.*, 1997). The glutathione-*S*-transferase (*GST1*) promoter of carnation flowers contains an 8 bp ERE sequence (ATTTCAAA) which conferred ethylene-regulated expression to a minimal *35S-CaMV* TATA-box promoter in an orientation-independent manner (Itzhaki *et al.*, 1994). Such an element is present in inverse orientation in the ozone-responsive *Vst1* promoter region (–408 to –401 bp) (*Figure 2*). A similar element (tTTTCAAA) is present at position –325 to –318. The *GST1* promoter of *Arabidopsis thaliana* contains also two repeated ERE elements (Yang *et al.*, 1998), and it has been shown that in addition to other stress factors, GST transcripts of *Arabidopsis thaliana* are also increased upon ozone fumigation (Conklin and Last, 1995; Sharma and Davis, 1997). Taken together this indicates that the ozone-responsive *Vst1* region is, at least in part, ethylene-regulated. Analysis of ozone-treated transgenic plants, harbouring either 5'-deletion constructs of the β-*1,3-glucanase* promoter (Vögeli-Lange *et al.*, 1994), or well-known mutation constructs in the GCC-box of the *PRB-1b* promoter (Sessa *et al.*, 1995) support the involvement of ethylene in signalling the ozone effect (B. Grimmig, G. Leubner-Metzger, R. Vögeli-Lange, F. Meins, G. Sessa, F. Fluhr, C. Langebartels, D. Ernst and H. Sandermann, unpublished data). However, for a basal ethylene-induced *Vst1* gene regulation sequences from –140 to –40 bp are sufficient (*Figure 1*) (B. Grimmig, N. Gonzalez, J. Penuelas, D. Ernst and H. Sandermann, unpublished data).

3.3 Salicylic acid- and jasmonic acid-responsive elements

SA and JA are important signals in several plant responses towards abiotic and biotic stress (Dong, 1998). SA is required for some, but not all ozone-induced mRNAs (Sharma *et al.*, 1996), and JA might also be involved in signalling ozone effects (Örvar *et al.*, 1997). It is well known that interactions between SA and JA signalling pathway(s) exist (Dong, 1998) and ozone might be interfering with both. The TCA-motif, present in a number of stress-induced genes, is an *in vitro*-binding site of a nuclear protein isolated from SA-treated tobacco (Goldsbrough *et al.*, 1993). Pairwise comparison of the ozone-responsive region with the TCA-motif showed no significant homology. In addition, a SA-responsive element (TTCGACC) of the tobacco β-*1,3-glu-canase* promoter (Durner *et al.*, 1997) is not present in the ozone-responsive region. However, this element is related to the W-boxes, present in promoters of several elicitor-induced genes (Rushton and Somssich, 1998), and similar motifs are present in the ozone-responsive *Vst1* region. JA-responsive domains contain a C-rich sequence and a G-box motif CACGTG(G), separated by a stretch of 26–29 bp (Mason *et al.*, 1993). In the ozone-responsive region two G-box-like elements (CcCGTGG, –342 bp; CAaGTGG, –314 bp) exist (*Figure 2*). The second G-box-like element is flanked by two C-rich domains at position –330 to –321 (CCTCCTTTTC) and –349 to –340 (CTTCC-

CTCCC). However, as G-boxes are involved in gene regulation of diverse environmental factors (Menkens *et al.*, 1995), they may also function upon ozone treatment, probably in combination with other *cis*-elements.

3.4 Elicitor-responsive elements

Elicitor-responsive elements (W-boxes) and structurally related cognate DNA-binding proteins have been characterized (Rushton *et al.*, 1996). The core hexamer sequence (T)TGAC(C) is sufficient for elicitor-induced *PR* gene expression in maize, parsley and tobacco (Rushton and Somssich, 1998). Two W-box-like core sequences are present in the ozone-responsive *Vst1* promoter region at position −388 (TGAC) and in inverse orientation at position −312 TTGAC (*Figure 2*). W-boxes are essential *cis*-elements interacting with the WRKY family of *trans*-factors and are present in several pathogen-responsive promoters (Rushton and Somssich, 1998). These genes are also induced by other stress factors, including ozone. The *Vst1* gene is induced by pathogens and ozone (Schubert *et al.*, 1997); therefore the presence of these two W-box-like elements may indicate that both stressors activate a signal pathway resulting in the binding of *trans*-factors to same *cis*-elements. However, the ozone-responsive *Vst1* region differs from the basal pathogen-responsive promoter region (Grimmig *et al.*, 1997; Schubert *et al.*, 1997). Therefore additional *cis*-elements must be involved in basic pathogen-inducible DNA-binding.

3.5 Additional stress-responsive element (STRE)

In yeast, a heat shock-independent stress sequence with a core consensus of AGGGG exists (Ruis and Schüller, 1995). This element is activated by multiple stress factors such as nitrogen starvation, pH changes, weak organic acids, ethanol, osmotic and oxidative stress. Such an element is also present at position −284 to −280 in the *Vst1* promoter (*Figure 2*). This element has been destroyed in the ozone-insensitive −280 bp deletion construct (*Figure 2*) (Schubert *et al.*, 1997). STREs are under negative control by protein kinase A (Ruis and Schüller, 1995), and there is strong evidence that phosphorylation cascades are involved in defence signalling of plants (Scheel, 1998). In tobacco an ozone-related ACC synthase phosphorylation/dephosphorylation has been described (Tuomainen *et al.*, 1997). Therefore, it is tentative to speculate that phosphorylation cascades are involved in ozone signalling.

Taking all these conserved sequence and motif identities, as well as similarities together, it is necesssary to remember that the analysis presented is based on computer-aided programs. Up to now no *in vivo* proof for the functionality of an 'ozone-responsive' element exists. First results performing electrophoretic mobility shift assays for the ozone-responsive *Vst1* region indicate no difference in the ability of proteins in nuclear extracts from control versus ozone-treated plants to bind to designed oligonucleotide probes (B. Grimmig, D. Ernst and H. Sandermann, unpublished data). This indicates that ozone-responsive *Vst1* gene induction is mediated by phosphorylation/dephosphorylation or by an interaction between multiple regulatory elements.

4. Biotechnological application

The introduction and successful expression of anti-microbial/-fungal genes in plants, resulting in enhanced resistance, has been demonstrated, and the use in field crops is anticipated (Salmeron and Vernooij, 1998). There is increasing evidence that for an efficient disease resistance based on gene expression in transgenic plants, a heterologous gene expression should be induced only after the plant–pathogen interaction had occurred (Fischer and Hain, 1994). The expression should be local and restricted to the infection site. Plants harbouring the *STS* gene under the control of the constitutive *35S-CaMV* promoter showed a lower degree of disease resistance compared to plants harbouring the *STS* gene under the control of its own *Vst1* promoter (Fischer and Hain, 1994). In general it is advantageous that an introduced resistance gene is not controlled by other stress factors, for example heavy metal ions, UV-B or ozone. The *Vst1* promoter is activated by ambient ozone concentrations, thus interfering with a pathogen-induced activation (Schubert *et al.*, 1997). Deletion of the *Vst1* promoter downstream to position –280 resulted in a basal pathogen-responsive region without ozone-inducibility (*Figure 1*) (Schubert *et al.*, 1997). By eliminating the ozone-responsive region it might be possible to design promoter constructs which become activated only by pathogens and not by ozone. Further stress-responsive regions have to be taken into account, because identical PR transcripts are induced by ozone, heavy metal ions and UV-B radiation in *Arabidopsis thaliana* (Heidenreich *et al.*, 1999).

Another strategy to enhance resistance is the development of chemicals that induce an inherent resistance response in plants (Gatz, 1997). SAR, a long-lasting disease resistance, can provide a significant level of protection against pathogens (Durner *et al.*, 1997). In many cases SAR is accompanied by the accumulation of endogenous SA. Exogenous application of SA or structural analogues such as aspirin, 2,6-dichloroisonicisotinic acid and benzothiadiazole (BTH) results in an induction of resistance against pathogens and the activation of SAR. BTH is currently developed as a commercial disease control chemical, and it has been shown that BTH induction of SAR is independent of SA, ethylene and jasmonate (Lawton *et al.*, 1996). BTH activates SAR either downstream of SA accumulation, or by acting as a SA analogue (Lawton *et al.*, 1996). Concerning the ozone-responsive *Vst1* region, the W-box-like elements may interact with SA analogues. Induction of W-box-containing defence genes by pathogens or elicitors can be potentiated by pretreatment with SA (Durner *et al.*, 1997). Therefore it might be possible that ambient ozone concentrations interfere with SA analogues, or act themselves as SA analogues.

Another group of chemicals, the safeners, are used to increase plant tolerance towards herbicides (Gatz, 1997). The mode of action is an enhanced metabolism of the herbicide applied. In addition to several other, well-known safener-inducible genes, specific GSTs transcripts are also induced upon safener application (Gatz, 1997). As GST mRNAs are also induced by ozone (Conklin and Last, 1995; Sharma and Davis, 1997), the detoxification effect of safeners may be influenced, even by typical ambient ozone concentrations.

In conclusion, the effect of ozone on gene regulation may interfere with several other stress-responsive promoter regions. This has to be taken into account in the use of gene induction by agrochemicals, as well as in the construction and transfer of defence-related genes and promoters.

Acknowledgements

The authors wish to thank their colleagues W. Heller, C. Langebartels and A. Schäffner for critically reading the manuscript. Financial support by EUROSILVA (BMBF), Fonds der Chemischen Industrie and Limagrain (Chappes, France) is gratefully acknowledged.

References

Abe, H., Yamaguchi-Shinozaki, K., Urao, T., Iwasaki, T., Hosokawa, D. and Shinozaki, K. (1997) Role of arabidopsis MYC and MYB homologs in drought- and abscisic acid-regulated gene expression. *Plant Cell* **9**: 1859–1868.
Betz, C. (1998) Rolle der Ethylenbiosynthese bei der Empfindlichkeit von Pflanzen gegenüber Ozon und oxidativem Stress. Thesis, LMU München.
Conklin, P.L. and Last, R.L. (1995) Differential accumulation of antioxidant mRNAs in *Arabidopsis thaliana* exposed to ozone. *Plant Physiol.* **109**: 203–212.
Dong, X.N. (1998) SA, JA, ethylene, and disease resistance in plants. *Curr. Opinion Plant Biol.* **1**: 316–323.
Douglas, C.J. (1996) Phenylpropanoid metabolism and lignin biosynthesis – from weeds to trees. *Trends Plant Sci.* **1**: 171–178.
Durner, J., Shah, J. and Klessig, D.F. (1997) Salicylic acid and disease resistance in plants. *Trends Plant Sci.* **2**: 266–274.
Eckey-Kaltenbach, H., Ernst, D., Heller, W. and Sandermann, H. (1994) Biochemical plant responses to ozone. IV. Cross-induction of defensive pathways in parsley (*Petroselinum crispum* L.) plants. *Plant Physiol.* **104**: 67–74.
Ernst, D., Schraudner, M., Langebartels, C. and Sandermann, H. (1992) Ozone-induced changes of mRNA levels of β-1,3-glucanase, chitinase and 'pathogenesis-related' protein 1b in tobacco plants. *Plant Mol. Biol.* **20**: 673–682.
Ernst, D., Bodemann, A., Schmelzer, E., Langebartels, C. and Sandermann, H. (1996) β-1,3-glucanase mRNA is locally, but not systemically induced in *Nicotiana tabacum* L. cv. BEL W3 after ozone fumigation. *J. Plant Physiol.* **148**: 215–221.
Eyal, Y., Meller, Y., Lev-Yadun, S. and Fluhr, R. (1993) A basic-type PR-1 promoter directs ethylene responsiveness, vascular and abscission zone-specific expression. *Plant J.* **4**: 225–234.
Fischer, R. and Hain, R. (1994) Plant disease resistance resulting from the expression of foreign phytoalexins. *Curr. Opinion Biotech.* **5**: 125–130.
Fluhr, R. (1998) Ethylene perception: from two-component signal transducers to gene induction. *Trends Plant Sci.* **3**: 141–146.
Gatz, C. (1997) Chemical control of gene expression. *Annu. Rev. Plant Physiol. Plant. Mol. Biol.* **48**: 89–108.
Goldsbrough, A.P., Albrecht, H. and Stratford, R. (1993) Salicylic acid-inducible binding of a tobacco nuclear protein to a 10 bp sequence which is highly conserved amongst stress-inducible genes. *Plant J.* **3**: 563–571.
Graf, T. (1992) Myb: a transcriptional activator linking proliferation and differentiation in hematopoietic cells. *Curr. Opinion Gen. Dev.* **2**: 249–255.
Grimmig, B., Schubert, R., Fischer, R., Hain, R., Schreier, P.H., Betz, C., Langebartels, C., Ernst, D. and Sandermann, H. (1997) Ozone- and ethylene-induced regulation of a grapevine resveratrol synthase promoter in transgenic tobacco. *Acta Physiol. Plant* **19**: 467–474.

Grotewold, E., Drummond, B.J., Bowen, B. and Peterson, T. (1994) The *myb*-homologous *P* gene controls phlobaphene pigmentation in maize floral organs by directly activating a flavonoid biosynthetic gene subset. *Cell* **76**: 543–553.

Heidenreich, B., Ernst, D. and Sandermann, H. (1999) Mercuric ion-induced specific gene expression in *Arabidopsis thaliana*. *Int. J. Phytoremediation* (in press).

Itzhaki, H., Maxson, J.M. and Woodson, W.R. (1994) An ethylene-responsive enhancer element is involved in the senescence-related expression of the carnation glutathione-*S*-transferase (*GST1*) gene. *Proc. Natl Acad. Sci. USA* **91**: 8925–8929.

Kangasjärvi, J., Talvinen, J., Utriainen, M. and Karjalainen, R. (1994) Plant defence systems induced by ozone. *Plant Cell Environ.* **17**: 783–794.

Krupa, S.V., Grünhage, L., Jäger, H.-L., Nosal, M., Manning, W.J., Legge, A.H. and Hanewald, K. (1995) Ambient ozone (O_3) and adverse crop response: a unified view of cause and effect. *Environ. Pollut.* **87**: 119–126.

Langebartels, C., Ernst, D., Heller, W., Lütz, C., Payer, H.-D. and Sandermann, H. (1997) Ozone responses of trees: results from controlled chamber exposures at the GSF phytotron. In: *Forest Decline and Ozone* (eds H. Sandermann, A.R. Wellburn, and R.L. Heath). Springer, Berlin, pp. 163–200.

Lawton, K.A., Friedrich, L., Hunt, M., Weymann, K., Delaney, T., Kessmann, H., Staub, T. and Ryals, J. (1996) Benzothiadiazole induces disease resistance in *Arabidopsis* by activation of the systemic acquired resistance signal transduction pathway. *Plant J.* **10**: 71–82.

Livne, B., Faktor, O., Zeitoune, S., Edelbaum, O. and Sela, I. (1997) TMV-induced expression of tobacco β-glucanase promoter activity is mediated by a single, inverted, GCC motif. *Plant Sci.* **130**: 159–169.

Mason, H.S., DeWald, D.B. and Mullet, J.E. (1993) Identification of a methyl jasmonate-responsive domain in the soybean *vspB* promoter. *Plant Cell* **5**: 241–251.

Menkens, A.E., Schindler, U. and Cashmore, A.R. (1995) The G-box: a ubiquitous regulatory element in plants bound by the GBF family of bZIP proteins. *Trends Biochem. Sci.* **20**: 506–510.

Örvar, B.L., McPherson, J. and Ellis, B.E. (1997) Pre-activating wounding response in tobacco prior to high-level ozone exposure prevents necrotic injury. *Plant J.* **11**: 203–212.

Pell, E.J., Schlagnhaufer, C.D. and Arteca, R.N. (1997) Ozone-induced oxidative stress: mechanisms of action and reaction. *Physiol. Plant.* **100**: 264–273.

Quandt, K., Frech, C., Karas, K., Wingender, E. and Werner, T. (1995) MatInd and MatInspector: new fast and versatile tools for detection of consensus matches in nucleotide sequence data. *Nucl. Acids Res.* **23**: 4878–4884.

Ruis, H. and Schüller, C. (1995) Stress signaling in yeast. *BioEssays* **17**: 959–965.

Rushton, P.J., Somssich, I.E. (1998) Transcriptional control of plant genes responsive to pathogens. *Curr. Opinion Plant Biol.* **1**: 311–315.

Rushton, P.J., Torres, J.T., Parniske, M., Wernert, P., Hahlbrock, K. and Somssich, I.E. (1996) Interaction of elicitor-induced DNA-binding proteins with elicitor response elements in the promoters of parsley PR1 genes. *EMBO J.* **15**: 5690–5700.

Salmeron, J.M. and Vernooij, B. (1998) Transgenic approaches to microbial disease resistance in crop plants. *Curr. Opinion Plant Biol.* **1**: 347–352.

Sandermann, H. (1998) Ozone: an air pollutant acting as a plant-signaling molecule. *Naturwissenschaften* **85**: 369–375.

Sandermann, H. (1996) Ozone and plant health. *Annu. Rev. Phytopathol.* **34**: 347–366.

Sandermann, H., Ernst, D., Heller, W. and Langebartels, C. (1998) Ozone: an abiotic elicitor of plant defence reactions. *Trends Plant Sci.* **3**: 47–50.

Scheel, D. (1998) Resistance response physiology and signal transduction. *Curr. Opinion Plant Biol.* **1**: 305–310.

Schraudner, M., Moeder, W., Wiese, C., Van Camp, W., Inzé, D., Langebartels, C. and Sandermann, H. (1998) Ozone-induced oxidative burst in the ozone biomonitor plant, tobacco Bel W3. *Plant J.* **16**: 235–245.

Schubert, R., Fischer, R., Hain, R., Schreier, P.H., Bahnweg, G., Ernst, D. and Sandermann, H. (1997) An ozone-responsive region of the grapevine resveratrol synthase promoter differs from the basal pathogen-responsive sequence. *Plant Mol. Biol.* **34**: 417–426.

Sessa, G., Meller, Y. and Fluhr, R. (1995) A GCC element and a G-box motif participate in ethylene-induced expression of the *PRB-1b* gene. *Plant Mol. Biol.* **28:** 145–153.

Sharma, Y.K. and Davis, K.R. (1997) The effects of ozone on antioxidant responses in plants. *Free Rad. Biol. Med.* **23:** 480–488.

Sharma, Y.K., León, J., Raskin, I. and Davis, K.R. (1996) Ozone-induced responses in *Arabidopsis thaliana*: the role of salicylic acid in the accumulation of defense-related transcripts and induced resistance. *Proc. Natl Acad. Sci. USA* **93:** 5099–5104.

Tuomainen, J., Betz, C., Kangasjärvi, J., Ernst, D., Zu-Hua, Y., Langebartels, C. and Sandermann, H. (1997) Ozone induction of ethylene emission in tomato plants: regulation by differential accumulation of transcripts for the biosynthetic enzymes. *Plant J.* **12:** 1151–1162.

Vögeli-Lange, R., Fründt, C., Hart, C.M., Nagy, F. and Meins, F. (1994) Developmental, hormonal, and pathogenesis-related regulation of the tobacco class I β-1,3-glucanase B promoter. *Plant Mol. Biol.* **25:** 299–311.

Willekens, H., Van Camp, W., Van Montagu, M., Inzé, D., Langebartels, C. and Sandermann, H. (1994) Ozone, sulfur dioxide, and ultraviolet B have similar effects on mRNA accumulation of antioxidant genes in *Nicotiana plumbaginifolia* L. *Plant Physiol.* **106:** 1007–1014.

Yang, K.-Y., Kim, E.-Y., Kim, C.-S., Guh, J.-O., Kim, K.-C., Cho, B.-H. (1998) Characterization of a glutathione S-transferase *ATGST 1* in *Arabidopsis thaliana*. *Plant Cell Rep.* **17:** 700–704.

Yang, Y. and Klessig, D.F. (1996) Isolation and characterization of tobacco mosaic virus-inducible *myb* oncogene homolog from tobacco. *Proc. Natl Acad. Sci. USA* **93:** 14972–14977.

Chapter 6

Signal molecules in ozone activation of stress proteins in plants

Wolfgang Moeder, Sabine Anegg, Gabriele Thomas, Christian Langebartels and Heinrich Sandermann Jr.

1. Background

The air pollutant ozone is a stress factor known to affect plant growth and performance, cause necrotic leaf lesions in certain plant species and cultivars and to induce chlorosis and accelerated leaf senescence in others (Pell *et al.*, 1997; Sandermann, 1996). The analysis of phytotoxic effects of ozone has been very much improved by comparisons between sensitive and tolerant plant cultivars, including radish, plantain, pea, bean, soybean, clover and tobacco as well as deciduous and coniferous tree species (Guzy and Heath, 1993; Wellburn and Wellburn, 1996). The ozone-sensitive model plants usually show characteristic symptoms of injury in controlled chamber studies at ozone concentrations and doses which are close to the limit values for air quality regulations.

The most effective biomonitor plants have been selections of ozone-sensitive and ozone-tolerant cigar wrapper varieties, Bel W3 and Bel B (Heggestad, 1991). Tobacco Bel W3 is a highly suitable biomonitor plant because damage can be unequivocally attributed to ozone, the plant is easy to cultivate in the field with the exception of cold regions, and the related cv. Bel B can be used as a control for non-specific effects. Based on this potential, we have analysed tobacco Bel W3 and Bel B over the past few years to characterize ozone actions and, in particular, the plant's reactions to ozone as a model for oxidative stress responses (Langebartels *et al.*, 1996b; Sandermann *et al.*, 1998). Methods used in these studies have recently been summarized by Langebartels *et al.* (1999a).

Ozone flux into plant leaves is a function of atmospheric ozone concentration at canopy level, the exchange properties of the atmosphere and the sink strength of the plants (Grünhage *et al.*, 1993). Stomatal conductance is the single most important variable governing the uptake of ozone at the leaf level (Reich, 1987). During recent years, numerous studies have investigated to what extent stomatal conductance determines the ozone sensitivity of sensitive and tolerant plant cultivars. It was basically found that plants with higher stomatal uptake show stronger effects, but additional parameters

Plant Responses to Environmental Stress, edited by M.F. Smallwood, C.M. Calvert and D.J. Bowles.

43

would have to be assumed to explain ozone tolerance of some of the cultivars (Guzy and Heath, 1993; Wellburn and Wellburn, 1996).

Ozone is readily absorbed in the aqueous layer surrounding mesophyll cells: the apoplastic fluid. Components of the apoplastic fluid are therefore of primary interest as reaction partners, and phenolic compounds, polyamines and especially ascorbate have been implicated as major factors in differential ozone sensitivity (Sharma and Davis, 1997). An ozone-sensitive *Arabidopsis* mutant has been characterized to be deficient in ascorbate biosynthesis (Conklin *et al.*, 1996). However, the subcellular localization of ascorbate has not yet been worked out. As will be shown in this chapter, not only antioxidant defence molecules and enzymes but also signal molecules occur in the apoplastic fluid of ozone-treated plants. These molecules mediate plant responses to pathogen infection, suggesting that ozone interferes with the respective signalling pathways.

Visible ozone injury in tobacco is not homogeneously distributed on middle-aged leaves, but occurs as spot-like necrotic lesions of 1–2 mm diameter, sharply separated from healthy parts of the tissue. In future, this specific pattern has to be analysed. One possible explanation may be that micro-heterogeneities of leaf areas and single cells occur in response to oxidative stress. In this context, it is interesting to note that various ozone reactions predominantly take place in the leaf veins and in periveinal regions (Langebartels *et al.*, 1999b). These foci may result from signals transported through phloem or xylem, or may represent primary sites of defence against pathogens using the 'vascular highways' for spreading within the plant.

2. Signalling pathways in ozone-exposed tobacco

Biochemical and physiological ozone responses of plants are largely due to specific activation or decrease of gene expression and resulting changes in the accumulation of translated proteins. An important finding of recent years was that ozone induces defence genes known from plant–pathogen interactions. Beginning the list with basic β-1,3-glucanase (PR2) and chitinase (PR3; Schraudner *et al.*, 1992), ozone-induced genes now comprise more than 40 genes from all lines of plant defence, including PR proteins, anti-oxidative systems, cell wall modifications and polyphenol metabolism (Langebartels *et al.*, 1999b). These findings convincingly characterize ozone response as an active process in plants. It was additionally found that defined isoforms of these genes were induced which typically also show up in hypersensitive reactions following pathogen attack (Hammond-Kosack and Jones, 1996) or during systemic acquired resistance (SAR; Sticher *et al.*, 1997).

The similarity of ozone and pathogen responses suggests that both stressors activate the same or overlapping signal pathways involved in defence activation and/or in damage amplification processes. This overlap may be based on changes of ion fluxes across the plasmalemma (Heath and Taylor, 1997), and a biphasic response in the concentration of cytosolic free calcium has recently been reported in ozone-treated *Arabidopsis* (Clayton *et al.*, 1999). In addition, reactive oxygen intermediates (ROI) may occur in both types of stress (Lamb and Dixon, 1997; Sandermann *et al.*, 1998; Sharma and

Davis, 1997). It has long been known that ozone is readily transformed into ROIs in the apoplastic fluid (Heath and Taylor, 1997). A recent study from our laboratory shows that ozone activates a biphasic oxidative burst of cellular origin in the sensitive tobacco Bel W3, while a monophasic response is found in tolerant cv. Bel B (Schraudner et al., 1998). Interestingly, there was no difference in the apoplastic accumulation of ROIs between sensitive and tolerant tobacco cultivars within the ozone exposure period. Also, anti-oxidant genes, specific isoforms of glutathione peroxidase and catalase (Schraudner et al., 1998; Willekens et al., 1994), were transiently activated at similar levels within this time period. A second and sustained peak of ROIs, consisting particularly of H_2O_2, was, however, only found in the ozone-sensitive tobacco Bel W3. It occurred during post-cultivation in pollutant-free air, demonstrating its plant origin. The biphasic oxidative burst showed striking similarities with that known from incompatible interactions of plants with (necrotrophic) pathogens (Lamb and Dixon, 1997). It thus seems that ozone-derived ROIs or other products activate the cellular production of ROIs in ozone-sensitive tobacco, thereby leading to amplification of the response and to cellular death in defined regions of the leaf.

It has been demonstrated that ozone (Schraudner et al., 1996) and UV-B treatments (Conconi et al., 1996) can also induce defence reactions via lipid-based signal pathways. Important steps in this signal chain are the activation of phospholipases, the release of linoleic and linolenic acid from cell membranes and the formation of defined lipid peroxides. These latter compounds are then transformed into a variety of products, including aldehydes, alcohols and, via the octadecanoid pathway, into the plant hormone jasmonate (Creelman and Mullet, 1997). Jasmonate, a C_{12} cyclopentanone compound, is a well-known inducer of defence reactions, e.g. proteinase inhibitors (Farmer et al., 1998). According to Örvar et al. (1997), mechanical wounding or application of jasmonate reduced ozone-induced leaf damage in tobacco Bel W3. Defined volatile lipid peroxidation products have been found following pathogen infection of bean (Croft et al., 1993) and ozone exposure of tobacco Bel W3 (Heiden et al., 1999). In the latter study, mainly C_6 alcohols and aldehydes with cis-3-hexenol as a major component were emitted with similar kinetics as peak II accumulation of ROIs (Schraudner et al., 1998). It remains to be seen whether jasmonate, lipid hydroperoxides or their transformation products play a role in the ozone activation of defence proteins and lesion development.

The gaseous hormone ethylene which is emitted as an early burst in response to pathogen infection and other stressors, constitutes an important mediator of ozone-induced responses and cell death. It has been shown in herbaceous plants and trees that ethylene emission is correlated with ozone sensitivity (Wellburn and Wellburn, 1996; S. Anegg and C. Langebartels, unpublished data). Ethylene emission is correlated with a transient increase in the level of its biosynthetic precursor, 1-aminocyclopropane-1-carboxylate (ACC). The induction of ethylene and ACC is very specific and comprises transcriptional activation of defined isoforms of the biosynthetic enzymes, ACC synthase and ACC oxidase (Schlagnhaufer et al., 1997; Tuomainen et al., 1997). It has recently been shown that ACC is transiently translocated into the apoplastic fluid where it could serve as a short-distance signal as well as a substrate of cell wall peroxidases or apoplastic ACC oxidase (*Figure 1*; Möder *et al.*, 1999). Occurrence of ACC oxidase at the outer side of the plasmalemma has been described for apple fruit tissue (Ramassamy *et al.*, 1998) while no information is currently available for leaves. Ethylene, applied externally, activates a variety of defence proteins, including basic glucanases and chitinases, cellulase, peroxidases, ripening genes as well as its own biosynthetic enzymes (Kieber, 1997). Correspondingly, ozone induced basic PR proteins in tobacco and *Arabidopsis* in an early phase (Schraudner *et al.*, 1992; Sharma and Davis, 1997). Ernst *et al.* (Chapter 7) summarize evidence for ethylene-responsive promotor regions of stilbene synthase, basic PR1b and β-1,3-glucanase. It is therefore highly probable that part of the activation of ozone-related proteins is mediated by ethylene.

Salicylic acid is involved as a signal molecule in local (hypersensitive response, HR) and systemic reactions of plants (Durner *et al.*, 1997). Its biosynthesis proceeds from *t*-cinnamic acid via benzoic acid which is transformed through an H_2O_2-inducible enzyme, benzoic acid 2-hydroxylase, into salicylic acid. External application of salicylic acid or its methyl ester, methyl salicylate, leads to an induction of acidic (class II) PR proteins (Durner *et al.*, 1997). It was shown that ozone treatment of *Arabidopsis* and tobacco results in a rapid and significant accumulation of free salicylic acid (Sharma and Davis, 1997). Levels of salicylic acid conjugates increased up to several days after the treatment. Preliminary experiments show that ozone induces a biphasic accumulation of salicylic acid in ozone-sensitive tobacco (peak I: free; peak II: conjugated salicylic acid) while only the first peak was induced, at 10-fold lower rates, in tolerant cv. Bel B (W. Möder and C. Langebartels, unpublished data). An emission of methyl salicylate from the sensitive cultivar Bel W3 was found 1 day after the exposure, when visible injury also developed (Heiden *et al.*, 1999). It is therefore assumed that ozone induction of PR1 (Sharma and Davis, 1997) and acidic isoforms of β-1,3-glucanase and chitinase (Langebartels *et al.*, 1999b) are mediated by salicylic acid.

In conclusion, ozone exposure activates several differentially regulated signal pathways (ethylene, ROIs and lipid peroxidation products, salicylic acid) in sensitive

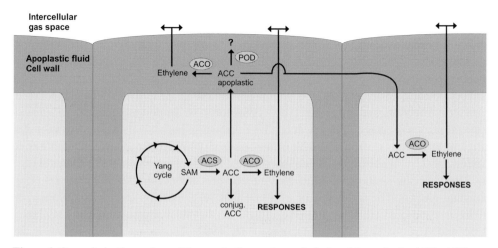

Figure 1. Ozone induction and possible apoplastic reactions of ethylene biosynthesis. ACO, ACC oxidase; ACS, ACC synthase; SAM, S-adenosyl methionine; POD, peroxidase.

tobacco and other plant species. Salicylic acid-independent defence pathways (Pieterse and Van Loon, 1999) seem to be induced in an early phase, while salicylic acid-dependent reactions (Durner *et al.*, 1997) are subsequently activated. As shown in *Figure 2*, activation of differentially localized signal molecules can be divided into three phases. In a first step, the ethylene precursor ACC, ROIs and lipid hydroperoxides, as well as salicylic acid are transiently accumulated intracellularly, with major differences in ethylene emissions and ACC as well as salicylic acid levels between sensitive and tolerant tobacco cultivars. In a second step, ACC, ROIs and free salicylic acid occur in the apoplastic fluid surrounding mesophyll cells. It may be speculated that these compounds act as short-distance signals in mediating responses to the neighbouring cells as proposed for ACC (Spanu and Boller, 1989) and H_2O_2 in plant–pathogen interactions (Lamb and Dixon, 1997). In addition, ACC might act as substrate of an apoplastic ACC oxidase leading to ethylene emission. The levels of all compounds are lower by factors of 10–50 in the apoplastic fluid of ozone-tolerant tobacco compared to the ozone-sensitive cultivar Bel W3. In a third phase, volatile signal molecules (ethylene, methyl salicylate) as well as C_6 alcohols and aldehydes (and H_2O_2?) are emitted into the intercellular gas space and can be detected in the atmosphere outside the leaf. Again, ozone-tolerant tobacco shows significantly lower emissions of all compounds compared to the sensitive one.

   It is not yet known whether the above compounds are formed in the same areas of the leaf that later transform into visible lesions. It is also largely unexplored whether these compounds act individually or in combination to induce the accumulation of defence proteins and enforce cellular damage in distinct leaf areas. ROIs, salicylic acid, cell death (and ethylene) have recently been proposed by Van Camp *et al.* (1998) to act in a self-amplifying 'oxidative cell death cycle'. Similar to the plant pathogen response, this cycle appears to act in ozone-sensitive tobacco as well. As a large range of reactions is activated by ozone or ozone-derived ROIs in tobacco Bel W3, it may be speculated that a regulatory step upstream of the above reactions is affected in this cultivar. This step may lead to hyperactivation in cv. Bel W3, while 'normal' oxidative stress recognition and responses are found in the tolerant counterpart. Targets of ozone-derived ROIs and

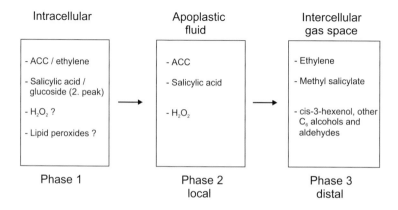

Intracellular	Apoplastic fluid	Intercellular gas space
- ACC / ethylene	- ACC	- Ethylene
- Salicylic acid / glucoside (2. peak)	- Salicylic acid	- Methyl salicylate
- H_2O_2 ?	- H_2O_2	- cis-3-hexenol, other C_6 alcohols and aldehydes
- Lipid peroxides ?		
Phase 1	Phase 2 local	Phase 3 distal

Figure 2. Induction and occurrence of signal molecules in ozone-treated plants.

of other primary products may be extracellular signal-regulated kinases ('ozone receptor' according to Sandermann (1996)) as well as other protein kinases or phosphatases involved in cellular H_2O_2 generation (Luan, 1998)). Future work with the 'model stress' ozone in tobacco will hopefully help to identify relevant sources of oxidant production and provide some understanding of localized leaf responses in response to oxidative stress.

Acknowledgements

Financial support by EUROSILVA (BMBF), DFG, Fonds der Chemischen Industrie and Limagrain (Chappes, France) is gratefully acknowledged.

References

Clayton, H., Knight, M.R., Knight, H., McAinsh, M.R. and Hetherington, A.M. (1999) Dissection of the ozone-induced calcium signature. *Plant J.* **17**: 575–579.
Conconi, A., Smerdon, M.J., Howe, G.A. and Ryan, C.A. (1996) The octadecanoid signalling pathway in plants mediates a response to ultraviolet radiation. *Nature* **383**: 826–829.
Conklin, P., Williams, E. and Last, R. (1996) Environmental stress sensitivity of an ascorbic acid-deficient *Arabidopsis* mutant. *Proc. Natl Acad. Sci. USA* **93**: 9970–9974.
Creelman, R.A. and Mullet, J.E. (1997) Biosynthesis and action of jasmonates in plants. *Annu. Rev. Plant Physiol. Plant Mol. Biol.* **48**: 355–381.
Croft, K.P.C., Jüttner, F. and Slusarenko, A.J. (1993) Volatile products of the lipoxygenase pathway evolved from *Phaseolus vulgaris* L. leaves inoculated with *Pseudomonas syringae* pv *phaseolicola*. *Plant Physiol.* **101**: 13–24.
Durner, J., Shah, J. and Klessig, D.F. (1997) Salicylic acid and disease resistance in plants. *Trends Plant Sci.* **2**: 266–274.
Farmer, E.E., Weber, H. and Vollenweider, S. (1998) Fatty acid signaling in *Arabidopsis*. *Planta* **206**: 167–174.
Grünhage, L., Dämmgen, U., Haenel, H.-D., Jäger, H.-J., Holl, A., Schmitt, J. and Hanewald, K. (1993) A new potential air-quality criterion derived from vertical flux densities of ozone and from plant response. *Angew Bot.* **67**: 9–13.
Guzy, M.R. and Heath, R.L. (1993) Responses to ozone of varieties of common bean (*Phaseolus vulgaris* L.). *New Phytol.* **124**: 617–625.
Hammond-Kosack, K.E. and Jones, J.D.G. (1996) Resistance gene-dependent plant defense responses. *Plant Cell* **8**: 1773–1791.
Heath, R.L. and Taylor, G.E. (1997) Physiological processes and plant responses to ozone exposure. In: *Ozone and Forest Decline: A Comparison of Controlled Chamber and Field Experiments* (eds H. Sandermann, A.R. Wellburn and R.L. Heath). Ecol Studies, Vol. 127. Springer, Berlin, pp. 317–368.
Heggestad, H.E. (1991) Origin of Bel-W3, Bel-C and Bel-B tobacco varieties and their use as indicators of ozone. *Environ. Pollut.* **74**: 264–291.
Heiden, A.C., Hoffmann, T. and Kahl, J. *et al.* (1999) Emission of volatile signal and defense molecules from ozone-exposed plants. *Ecol. Appl.* (in press).
Kieber, J.J. (1997) The ethylene response pathway in *Arabidopsis*. *Annu. Rev. Plant Physiol. Plant Mol. Biol.* **48**: 277–296.
Lamb, C. and Dixon, R.A. (1997) The oxidative burst in plant disease resistance. *Annu. Rev. Plant Physiol. Plant Mol. Biol.* **48**: 251–275.
Langebartels, C., Ernst, D., Kangasjärvi, J. and Sandermann, H. (1999a) Ozone effects on plant defense. *Meth. Enzymol.* (in press).
Langebartels, C., Schraudner, M., Heller, W., Ernst, D. and Sandermann, H. (1999b) Oxidative

stress and defense reactions in plants exposed to air pollutants and UV-B radiation. In: *Oxidative stress in plants* (eds D. Inzé and M. Van Montagu). Harwood, Amsterdam (in press).

Luan, S. (1998) Protein phosphatases and signaling cascades in higher plants. *Trends Plant Sci.* **3**: 271–275.

Möder, W., Kangasjärvi, J., Elstner, E.F., Langebartels, C. and Sandermann, H. (1999) Apoplastic ACC in ozone- and elicitor-treated plants. In: *Biology and Biotechnology of the Plant Hormone Ethylene II* (eds A.K. Kanellis, C. Chang, H. Kende and D. Grierson). Kluwer, Dordrecht (in press).

Örvar, B.L., McPherson, J. and Ellis, B.E. (1997) Pre-activating wounding response in tobacco prior to high-level ozone exposure prevents necrotic injury. *Plant J.* **11**: 203–212.

Pell, E.J., Schlagnhaufer, C.D. and Arteca, R.N. (1997) Ozone-induced oxidative stress: mechanisms of action and reaction. *Physiol. Plant* **100**: 264–273.

Pieterse, C.M.J. and Van Loon, L.C. (1999) Salicylic acid-independent plant defence pathways. *Trends Plant Sci.* **4**: 52–58.

Ramassamy, S., Olmos, E., Bouzayen, M., Pech, J.C. and Latché, A. (1998) 1-Aminocyclopropane-1-carboxylate oxidase of apple fruit is periplasmic. *J. Exp. Bot.* **49**: 1909–1915.

Reich, P.B. (1987) Quantifying plant response to ozone: a unifying theory. *Tree Physiol.* **3**: 63–91.

Sandermann, H. (1996) Ozone and plant health. *Annu. Rev. Phytopathol.* **34**: 347–366.

Sandermann, H., Ernst, D., Heller, W. and Langebartels, C. (1998) Ozone: an abiotic elicitor of plant defense reactions. *Trends Plant Sci.* **3**: 47–50.

Schlagnhaufer, C.D., Arteca, R.N. and Pell, E.J. (1997) Sequential expression of two 1-aminocyclopropane-1-carboxylate synthase genes in response to biotic and abiotic stresses in potato (*Solanum tuberosum* L.) leaves. *Plant Mol. Biol.* **35**: 683–688.

Schraudner, M., Ernst, D., Langebartels, C. and Sandermann, H. (1992) Biochemical plant responses to ozone. III. Activation of the defense-related proteins β-1,3-glucanase and chitinase in tobacco leaves. *Plant Physiol.* **99**: 1321–1328.

Schraudner, M., Langebartels, C. and Sandermann, H. (1996) Plant defence systems and ozone. *Biochem. Soc. Trans.* **24**: 456–461.

Schraudner, M., Möder, W., Wiese, C., Van Camp, W., Inzé, D., Langebartels, C. and Sandermann, H. (1998) Ozone-induced oxidative burst in the ozone biomonitor plant, tobacco Bel W3. *Plant J.* **16**: 235–245.

Sharma, Y.K. and Davis, K.R. (1997) The effects of ozone on antioxidant responses in plants. *Free Radical Biol. Med.* **23**: 480–488.

Spanu, P. and Boller, T. (1989) Ethylene biosynthesis in tomato plants infected by *Phytophthora infestans*. *J. Plant Physiol.* **134**: 533–537.

Sticher, L., Mauch-Mani, B. and Métraux, J.P. (1997) Systemic acquired resistance. *Annu. Rev. Phytopathol.* **35**: 235–270.

Tuomainen, J., Betz, C., Kangasjärvi, J., Ernst, D., Yin, Z.-H., Langebartels, C. and Sandermann, H. (1997) Ozone induction of ethylene emission in tomato plants: regulation by differential accumulation of transcripts for the biosynthetic enzymes. *Plant J.* **12**: 1151–1162.

Van Camp, W., Van Montagu, M. and Inzé, D. (1998) H_2O_2 and NO: redox signals in disease resistance. *Trends Plant Sci.* **3**: 330–334.

Wellburn, F.A.M. and Wellburn, A.R. (1996) Variable patterns of antioxidant protection but similar ethene emission differences in several ozone-sensitive and ozone-tolerant plant selections. *Plant Cell Environ.* **19**: 754–760.

Willekens, H., Van Camp, W., Van Montagu, M., Inzé, D., Langebartels, C. and Sandermann, H. (1994) Ozone, sulfur dioxide, and ultraviolet B have similar effects on mRNA accumulation of antioxidant genes in *Nicotiana plumbaginifolia* L. *Plant Physiol.* **106**: 1007–1014.

Chapter 7

Iron and oxidative stress in plants

Pierre Fourcroy

1. Background

Oxygen and iron are both essential and noxious for living organisms. The main form of life on earth (aerobic life) is conditioned by the availability of oxygen. However, oxygen metabolism gives rise to molecular species that can be toxic, even lethal, for living organisms including bacteria, yeast, plants and animals. The damage to proteins, membrane lipids and DNA caused by the accumulation of reduced oxygen intermediates result in what is usually described as an oxidative stress. The most reactive oxygen species (ROS) is the hydroxyl radical which can be produced in the presence of heavy metals such as Fe, Cu, Ni or Zn. Thus, heavy metals potentiate the toxicity of ROS. Cells of all organisms, from bacteria to mammals, possess systems able to eliminate toxic oxygen species and to regulate iron homeostasis. In *Arabidopsis thaliana* at least 15 proteins in charge of scavenging toxic oxygen species have been identified and characterized at the molecular level: superoxide dismutases (SODs), catalases, ascorbate peroxidases (APXs) and glutathione peroxidases. Most of these enzymes are present as different isoforms with specific subcellular localization. This fact could be interpreted as an absolute need for the plants to prevent damage caused by ROS in all cellular compartments. Yet, little is known on the specific role of the individual defence protein. Do they all participate in response to all stresses? Is there a hierarchical order in environmental stresses? Will transgenic plants possessing higher antioxidant defences be more tolerant to high salinity or drought?

2. Key questions

Our current knowledge of the relation between iron and oxidative stress needs to be extended. Concerning the effect of an excess of iron in plants, almost all attention has focused up to now on ferritin gene expression. Indeed ferritin can be considered as a defence protein since it sequesters the iron in a non-toxic form. The fact that another gene, that is APX, can be also up-regulated by iron may be an indication that other plant defences might be elicited by this metal. Although they are mechanisms controlling iron homeostasis (uptake, storage), stress conditions may temporarily affect the level of free iron. Little is known about the re-cycling of iron from plant metalloproteins *in vivo*. The haem oxygenase, which is a key enzyme in recycling iron from haemoglobin in animals, has not yet been cloned from higher plants. Another difficulty is that it is not easy to determine the relative amount of free iron and that which is bound to proteins.

3. Plant defence against reactive oxygen species

Plants, like all aerobic organisms, are exposed to ROS which are by-products of oxygen consumption. ROS are responsible for protein, lipid and nucleic acid degradation and are thought to play a major role in ageing and cell death (Jacobson, 1996). ROS are normal products of the metabolism but they can accumulate in the case of various environmental stress conditions. To avoid the accumulation of these compounds to toxic level, animals and plants possess several detoxifying enzymatic systems. Two main classes of plant defences have been described and can be classified as non-enzymatic and enzymatic systems. The first class (non-enzymatic) consists of small molecules such as ascorbate, glutathione, α-tocopherol which can react directly with the ROS. Enzymatic defences have the capacity to eliminate superoxides (SODs) or peroxides (catalases, peroxidases).

While SODs and catalases are found in animals, plants and microorganisms, APX (EC 1.11.1.11) is a plant-specific hydrogen peroxide-scavenging enzyme (Asada, 1992) which was first demonstrated to be active in the chloroplast in the so-called Halliwell–Asada pathway (*Figure 1*). APX isoforms are encoded by a nuclear multigene family and are found in various subcellular compartments. APX has been detected in both the cytosol and chloroplasts (Asada, 1992). More recently a pumpkin peroxisomal (Yamaguschi *et al.*, 1995) and a cotton glyoxysomal (Bunkelmann and Trelease, 1996) APX have been isolated.

The activities of antioxidant enzymes such as APX catalases and SODs are upregulated in response to several abiotic stresses such as drought (Smirnoff and

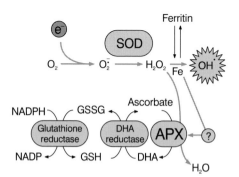

Figure 1. The Halliwell–Asada pathway (ascorbate–glutathione cycle). An excess of electrons produced during photosynthesis can react with molecular oxygen to produce superoxide. Superoxide dismutase (SOD) will produce hydrogen peroxide that will be eliminated by ascorbate peroxidase (APX). Ascorbate, the electron donor, will be regenerated through redox cycles involving glutathione (GSH) and NADP. Hydrogen peroxide can lead to the synthesis of the highly toxic hydroxyl radical in the presence of heavy metals such as Fe.

Colombe, 1988), low temperature (Schöner and Krause, 1990), high light intensities (Camak and Marschner, 1992), ozone, SO_2, UV-B (Willekens et al., 1994) and salinity (Lopez et al., 1996). All these adverse environmental conditions have a common consequence, that is the generation of an oxidative stress in plants.

4. Iron plays a dual role in cellular metabolism

As a constituent of metalloproteins (e.g. cytochromes), iron is essential for enzymatic reactions involving electron transfer. In contrast, iron can catalyse the production of the highly reactive hydroxyl radicals from hydrogen peroxide through the Fenton reaction. It is therefore important to avoid the occurrence of a large pool of free iron. To control the level of free iron, animal and plant cells sequester iron in specialized structures such as ferritins. The expression of ferritin genes is controlled by iron in both animals and plants, however while the main level of control is translational in animals, it is transcriptional in plants (Lescure et al., 1991), resulting in the accumulation of ferritin mRNA in response to an excess of iron (Lobréaux et al., 1995).

5. Iron and oxygen metabolisms are linked

Recently, increasing attention has focused on the integration of oxygen and iron metabolism (Rouault and Klausner, 1996). Indirect evidence for the inter-dependence of iron homeostasis and oxidative balance has been obtained in bacteria, yeast and mammals. For example, in *Escherichia coli*, the genes coding for Fe uptake and two superoxide dismutases (MnSOD and FeSOD) are regulated by the same transcription factors (Fee, 1991). In the yeast *Saccharomyces cerevisiae*, the transcription factor MAC1 is required for both the expression of a Fe^{3+} reductase and a cytosolic catalase (Jungmann et al., 1993). In mammalian cells, hydrogen peroxide is able to activate the regulator IRP1, responsible for the post-transcriptional control of the synthesis of proteins involved in iron homeostasis (Pantopoulos and Hentze, 1995).

In plants exposed to an excess of iron, antioxidant molecules such as glutathione and ascorbate are converted to their oxidized forms (Kampfenkel et al., 1995). Moreover, iron-loading of *Nicotiana plumbaginifolia* leaves led to an increase in catalase and APX activities (Kampfenkel et al., 1995). However, the only documented example of plant gene expression activated by iron through a major transcriptional control is ferritin, an iron storage protein localized within the plastids. In maize, ferritin mRNA accumulation is induced by iron (Lobréaux et al., 1992) as a result of an oxidative stress (Lobréaux et al., 1995).

6. Iron regulates the expression of ascorbate peroxidase

Do plants raise their antioxidant defences in response to an increase in free iron? We attempted to answer this question by selecting APX as a putative candidate for the following reasons. Firstly, APX is an iron-containing protein (Asada, 1992), therefore iron availability could regulate the expression of apx gene(s). Indeed, an iron-deficiency in *Euglena gracilis* has been shown to result in the complete loss of APX activity (Ishikawa *et al.*, 1993). Secondly, iron and ascorbate metabolisms are intimately linked, because an iron-mediated oxidative stress triggers both ferritin accumulation (Lobréaux *et al.*, 1995) and ascorbate oxidation to dehydroascorbate (Kampfenkel *et al.*, 1995), a condition favouring iron uptake by plant ferritin (Laulhère and Briat, 1993; Laulhère *et al.*, 1990). Thirdly, a protein sharing 71% homology in its N-terminal sequence with the cytosolic form of APX from *Arabidopsis thaliana* is accumulated in the tomato *chloronerva* mutant which accumulates abnormally high levels of copper and iron (Herbik *et al.*, 1996).

In dicots the uptake of iron implies several steps involving the reduction of Fe^{3+} to Fe^{2+}. A convenient way to allow a rapid iron uptake is to cut off most of the root system so that the iron loading of leaves can occur through the transpiration flux. We then used de-rooted rapeseed seedlings (6 days old) as an experimental system. No visible symptoms of injury (necrotic spots) were detectable in *Brassica napus* seedlings treated in this way during the time course of experiments, that is up to 24 h.

When *B. napus* seedlings were exposed to a solution of Fe(III)-citrate (500 mM), a rapid and large increase of the *apx* mRNA level occurred. The triggering of *apx* expression by iron was also clearly demonstrated by using increasing amounts of Fe(III)-citrate: a progressive accumulation of *apx* mRNA was recorded when the iron concentration was raised from 50 to 500 μM (*Figure 2*). Furthermore, the *apx* mRNA accumulation was shown to be iron-specific (Vansuyt *et al.*, 1997). Transition or heavy metals such as Zn, Ni or Co had no effect on the *apx* gene expression. Copper induced a slight increase in *apx* transcripts at the relatively high concentration of 5 mM, that is 10-fold higher than that of Fe(III)-citrate which induces a massive accumulation of *apx* mRNA.

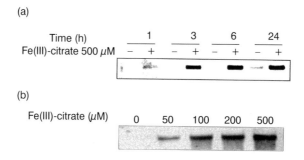

Figure 2. Iron-induced *apx* mRNA accumulation in *Brassica napus* cotyledons. *B. napus* seedlings were grown in the absence of iron; the roots were excised and the plantlets were transferred to water containing iron at the indicated concentrations. RNA was extracted and subjected to northern blot analysis with either the *apx* or *rbcS* probes. (a) Time course induction of *apx* mRNA accumulation. (b) Effect of increasing Fe(III)-citrate concentrations for 3 h. Modified from *FEBS Letters*, vol. 410, Vansuyt G. *et al.*, Iron triggers a rapid induction of ascorbate peroxidase gene expression in *Brassica napus*, pp. 195–200, Copyright 1997, with permission from Elsevier Science.

7. Iron and oxidative stress

The best understood cause for iron toxicity is its ability to catalyse the formation of the very toxic hydroxyl radical from superoxide and hydrogen peroxide. It has been proposed that an iron-mediated oxidative stress triggers the activation of some ferritin genes (Lobréaux et al., 1995). Since it is important for the cells to scavenge hydrogen peroxide, one could hypothesize that H_2O_2 would induce the expression of APX. No accumulation of apx mRNA was recorded in response to hydrogen peroxide. The treatment of B. napus seedlings with the thiol-oxidizing compound diamide, used at 1.5 mM, also had no effect on apx expression (Vansuyt et al., 1997). Although both H_2O_2 and diamide are not inducers of apx mRNA accumulation, one cannot exclude that other pro-oxidants might regulate the apx gene expression. Such a situation has already been observed in yeast: the expression of yap1 and trx2 are strongly activated by H_2O_2 and diamide, but not by t-BOOH or paraquat (Kuge and Jones, 1994).

We did not observe any stimulation of apx mRNA accumulation in B. napus seedlings treated with antioxidants such as N-acetyl cysteine (NAC), glutathione or ascorbate. Using a transient expression system in tobacco protoplasts, it has been shown that the promoter of the cytosolic Cu/Zn sod was inducible by reduced glutathione, dithiothreitol and NAC above 1 mM (Hérouart et al., 1993). Our data show that under our experimental conditions apx expression is not activated by reducing agents suggesting that the apx promoter possesses control elements different to those present in the sod promoter. In addition, antioxidants such as NAC failed to inhibit the iron activation of apx gene(s) (Vansuyt et al., 1997). Thus, the signalling pathway of apx expression triggered by iron differs from that of ferritin since the iron induction of ferritin synthesis is abolished by NAC in B. napus seedlings as well as in maize (Lobréaux et al., 1995).

8. Specificity of the iron induction of gene expression

The fact that APX is encoded by a multigene family raises the question of which apx genes are regulated by iron. Using probes corresponding to the 3' non-coding region of three different apx cDNA (encoding the cytosolic form APX1, a peroxisomal and a chloroplastic form) from A. thaliana, we reached the conclusion that the one cytosolic apx gene was inducible by iron. Further experiments are, however, needed to investigate the possible induction of the other apx forms. The promoter of the cytosolic apx1 has been fused with the gus coding sequence. First analysis of the transgenic A. thaliana obtained confirmed that this promoter is iron-inducible. In addition, attempts to identify other antioxidant defence genes responding to iron have failed up to now.

9. Future directions

We have shown that APX gene expression can be controlled by iron at the level of mRNA accumulation. Until now, ferritin was the only known plant gene with an iron-mediated expression. The fact that the expression of APX, an enzyme participating in the scavenging of ROS, is iron-dependent provides an attractive model with which to decipher the molecular links between transition metal metabolism and oxidative stress in plants. Another novel aspect of this work is that the accumulation of *apx* mRNA, although sharing some kinetics and dose-responsive iron-induced gene expression with ferritin, appears to be regulated by a specific pathway which remains to be elucidated. Iron-inducibility of both ferritin and APX genes suggest the occurrence of an integrated system. After sensing an excess of free iron, plant cells will react by decreasing both free iron and hydrogen peroxide, thus avoiding the accumulation of the highly toxic hydroxyl radical.

Analysing the mechanisms underlying the induction of APX by iron provides a convenient model for studying the molecular links between transition metal metabolism and oxidative stress in plants. The iron-responsive sequences of the gene coding for one cytosolic form of *A. thaliana* APX will be localized using a set of deletions of the promoter fused to the *gus* gene. Sequence analysis of ferritin and APX promoters may allow us to find common control elements. Identifying other plant genes that can be iron-responsive could help us to understand how plants react to changes in cellular iron status.

Acknowledgements

Transgenic plants expressing the *apx1* promoter–*gus* fusion were kindly provided by D. Inzé and S. Kushnir.

References

Asada, K. (1992) Ascorbate peroxidase – a hydrogen-scavenging enzyme in plants. *Physiol. Plant.* **85:** 235–241.

Bunkelmann, J.R. and Trelease, R.N. (1996) Ascorbate peroxidase. A prominent membrane protein in oilseed glyoxysomes. *Plant Physiol.* **110:** 589–598.

Camak, I. and Marschner, H. (1992) Magnesium deficiency and high light intensity enhance activities of superoxide dismutase, ascorbate peroxidase, and glutathione reductase in bean leaves. *Plant Physiol.* **98:** 1222–1227.

Fee, J.A. (1991) Regulation of *sod* genes in *Escherichia coli*: relevance to superoxide dismutase function. *Mol. Microbiol.* **5:** 2599–2610.

Herbik, A., Giritch, A., Horstmann, C., Becker, R., Balzer, H.J., Bäumlein, H. and Stephan, U.W. (1996) Iron and copper nutrition-dependent changes in protein expression in a tomato wild type and the nicotianamine-free mutant *chloronerva*. *Plant Physiol.* **111:** 533–540.

Hérouart, D., Van Montagu, M. and Inzé, D. (1993) Redox-activated expression of the cytosolic copper/zinc superoxide dismutase gene in *Nicotiana*. *Proc. Natl Acad. Sci. USA* **90:** 3108–3112.

Ishikawa, T., Takeda, T., Shigeoka, S., Hirayama, O. and Mitsunaga, T. (1993) Requirement for iron and its effect on ascorbate peroxidase in *Euglena gracillis*. *Plant Sci.* **93:** 25–29.

Jacobson, M.D. (1996) Reactive oxygen species and programmed cell death. *Trends Biochem. Sci.* **21:** 83–86.

Jungmann, J.H.-A.R., Lee, J., Romeo, A., Hassett, R., Kosman, D. and Jentsch, S. (1993) MAC1, a nuclear regulatory protein related to Cu-dependent transcription factors is involved in Cu/fe utilization and stress resistance in yeast. *EMBO J.* **12:** 5051–5056.

Kampfenkel, K., Van Montagu, M. and Inzé, D. (1995) Effects of iron excess on *Nicotiana plumbaginifolia* plants. *Plant Physiol.* **107:** 725–735.

Kuge, S. and Jones, N. (1994) YAP1 dependent activation of TRX2 is essential for the response of *Saccharomyces cerevisiae* to oxidative stress by hydroperoxides. *EMBO J.* **13**: 655–664.

Laulhère, J.P. and Briat, J.F. (1993) Iron release and uptake by plant ferritin is affected by pH, reduction and chelation. *Biochem. J.* **290**: 693–699.

Laulhère, J.P., Labouré, A.M. and Briat, J.F. (1990) Photoreduction and incorporation of iron into ferritins. *Biochem. J.* **269**: 79–84.

Lescure, A.M., Proudhon, D., Pesey, H., Ragland, M., Theil, E.C. and Briat, J.F. (1991) Ferritin gene transcription is regulated by iron in soybean cell cultures. *Proc. Natl Acad. Sci. USA* **88**: 8222–8226.

Lobréaux, S., Massenet, O. and Briat, J.F. (1992) Iron induces ferritin synthesis in maize plantlets. *Plant Mol. Biol.* **19**: 563–575.

Lobréaux, S., Thoiron, S. and Briat, J.F. (1995) Induction of ferritin synthesis in maize leaves by an iron-mediated oxidative stress. *Plant J.* **8**: 443–449.

Lopez, F., Vansuyt, G., Casse-Delbart, F. and Fourcroy, P. (1996) Ascorbate peroxidase activity, not mRNA level is enhanced in salt-stressed *Raphanus sativus* plants. *Physiol. Plant.* **97**: 13–20.

Pantopoulos, K. and Hentze, M.W. (1995) Rapid responses to oxidative stress mediated by iron regulatory protein. *EMBO J.* **14**: 2917–2924.

Rouault, T.A. and Klausner, R.D. (1996) Iron-sulfur clusters as biosensors of oxidants and iron. *Trends Biochem. Sci.* **21**: 174–177.

Schöner, S. and Krause, G.H. (1990) Protective systems against active oxygen species in spinach: response to cold acclimation in excess light. *Planta* **180**: 383–389.

Smirnoff, N. and Colombe, S.V. (1988) Drought influences the activity of enzymes of the chloroplast hydrogen peroxide scavenging system. *J. Exp. Bot.* **39**: 1097–1108.

Vansuyt, G., Lopez, F., Inzé, D., Briat, J.F. and Fourcroy, P. (1997) Iron triggers a rapid induction of ascorbate peroxidase gene expression in *Brassica napus*. *FEBS Lett.* **410**: 195–200.

Willekens, H., Van Camp, W., Van Montagu, M., Inzé, D., Langebartels, C. and Sandermann, H. (1994) Ozone, sulfur dioxide, and ultraviolet B have similar effects on mRNA accumulation of antioxidant genes in *Nicotiana plumbaginifolia* L. *Plant Physiol.* **106**: 1007–1014.

Yamaguschi, K., Mori, H. and Nishimura, M. (1995) A novel isoenzyme of ascorbate peroxidase localized on glyoxysomal and leaf peroxisomal membranes in pumpkin. *Plant Cell Physiol.* **36**: 1157–1162.

Chapter 8

H_2O_2 signalling in plant cells

Steven Neill, Radhika Desikan, Andrew Clarke and John Hancock

1. Background

The reactive oxygen species (ROS) H_2O_2 arises in plant cells via a number of routes, including photorespiration in peroxisomes and via the action of H_2O_2-generating oxidases. Most of the H_2O_2 in cells is formed via the dismutation of superoxide (O_2^-), which arises as a result of single electron transfer to molecular oxygen in electron transport chains, principally during the Mehler reaction in chloroplasts (Bartosz, 1997). A number of abiotic stresses lead to increased amounts of ROS, in particular H_2O_2, within cells, with the consequent imposition of oxidative stress (Smirnoff, 1998). Such stresses include air pollution, high and low temperature, dehydration and UV radiation (Bartosz, 1997). It is also clear that biotic stress in the form of pathogen challenge similarly leads to elevated levels of H_2O_2 generated at the cell surface, in what is termed the oxidative burst (Wojtaszek, 1997). Consequently, key questions concern the mechanisms and location of ROS production under various stresses.

Rapid non-specific reactions of ROS include lipid peroxidation and damage to proteins and DNA with consequent necrosis (Bartosz, 1997). However, it is clear that H_2O_2 can also act as a signalling molecule mediating the acquisition of tolerance to both abiotic and biotic stresses (Foyer et al., 1997; Smirnoff, 1998). For example, exogenous H_2O_2 can induce tolerance to chilling, high temperature and biotic stress, all of which cause elevated endogenous H_2O_2 production (Foyer et al., 1997; Prasad et al., 1994; Vallelian-Bindeschedler et al., 1998). There are several reports of cross-tolerance and some cellular responses have been found to be common to a number of different stresses (Foyer et al., 1997). For example, a number of biotic and abiotic stresses can induce the expression of genes encoding antioxidant enzymes or proteins involved in defensive responses. Salicylic acid also seems to be a common feature of H_2O_2 signalling (e.g. Alvarez and Lamb, 1997; Sharma et al., 1996). It is possible that different stresses induce both common and distinct cellular responses, the particular spectrum of responses determining the output. For example, H_2O_2 may be generated during several stresses but may only reach threshold concentrations in certain situations, or may synergize with other molecules such as salicylic acid or nitric oxide to exert its effects (Dangl, 1998). The key requirements here are to identify the signalling pathways induced by H_2O_2 and to determine how they effect acclimation to various stresses.

Plant Responses to Environmental Stress, edited by M.F. Smallwood, C.M. Calvert and D.J. Bowles.
© 1999 BIOS Scientific Publishers, Oxford.

2. H_2O_2 signalling in suspension cultures of *Arabidopsis thaliana*

The oxidative burst, during which large amounts of H_2O_2 are generated at the cell surface, is now well established as one of the earliest responses of plant cells to attempted invasion by potentially phytopathogenic microorganisms or in response to challenge by various elicitor molecules. H_2O_2 generated during this oxidative burst has several effects: it may be directly microbicidal, it can mediate the oxidative cross-linking of cell wall polymers, it induces the expression of genes encoding proteins involved in defensive and antioxidant processes and it can induce programmed cell death characteristic of the hypersensitive response (Alvarez and Lamb, 1997; Wojtaszek, 1997). As discussed previously, there is increasing evidence that an oxidative burst can also be induced by various abiotic stresses such as UV radiation and heat stress and that abiotic stresses such as low temperature, dehydration and ozone exposure may also result in elevated H_2O_2 production (Doke, 1997; Foyer *et al.*, 1997; Smirnoff, 1998; Vallelian-Bindeshedler *et al.*, 1998). The oxidative burst can be induced both in plants and in suspension-cultured cells. Whilst suspension cultures can in no way mimic the intricacies of whole plant responses, they do represent a very useful system with which to dissect those signalling processes required for both the generation of H_2O_2 and the subsequent cellular responses that it induces.

We have been using suspension cultures of *Arabidopsis thaliana* as a model system for this purpose. Treatment of such cultures with harpin, a proteinaceous bacterial elicitor, induces a rapid oxidative burst that requires both protein phosphorylation and calcium influx (Desikan *et al.*, 1996, 1997). The source of the H_2O_2 generated during the oxidative burst is still not resolved and there may well be several (Wojtaszek, 1997). However, the pioneering work of Doke (see Doke, 1997) suggested that the H_2O_2 arises from the dismutation of O_2^-, which is formed via the single electron reduction of molecular oxygen catalysed by a plasma membrane-located enzyme similar to the NADPH oxidase complex found in mammalian neutrophils. Since Doke's original work, a substantial body of evidence has accumulated to support the existence of a plasma membrane NADPH oxidase complex in plants. In *Arabidopsis* suspension cultures, pharmacological and biochemical evidence is consistent with the activity of NADPH oxidase (Desikan *et al.*, 1996), and homologues of *gp91*, the key redox component of the enzyme complex, have been cloned from *Arabidopsis* (Desikan *et al.*, 1998b; Keller *et al.*, 1998; Torres *et al.*, 1998).

In addition to the oxidative burst, expression of defence-related genes is also induced by harpin in *Arabidopsis* suspension cultures. Such genes include *PAL*, encoding phenylalanine ammonia lyase, a key enzyme of phenylpropanoid metabolism and *GST*, encoding glutathione-*S*-transferase, required for the detoxification of lipid hydroperoxides generated during oxidative stress, as well as the *gp91* homologue. The expression of these genes can also be induced by H_2O_2 in a time- and dose-dependent manner (*Figure 1*; Desikan *et al.*, 1998a, 1998b). It is difficult to calculate physiologically relevant concentrations of exogenous H_2O_2 as it is very rapidly degraded by plant cells (Bestwick *et al.*, 1997; Levine *et al.*, 1994), including *Arabidopsis* (Desikan *et al.*, 1998a), suggesting that apparently high amounts of externally added H_2O_2 might actually result in much

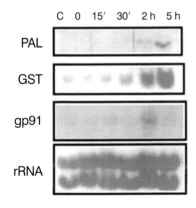

Figure 1. H_2O_2-induced gene expression in suspension-cultured cells of *Arabidopsis thaliana*. Cells were exposed to 10 mM H_2O_2 and RNA extracted at increasing time intervals and subjected to sequential northern analysis using as hybridization probes *PAL, GST* or *gp91*. Equal loadings were confirmed by subsequent hybridization with a rRNA probe. C, mock-treated cells.

lower concentrations at its site(s) of action. The mechanisms by which degradation of exogenous H_2O_2 are effected have not yet been resolved, although antioxidant capacity is associated with living cells and not culture medium (A. Clarke, unpublished data). Ascorbate peroxidase and catalase represent the major enzymes of H_2O_2 degradation, and it is interesting to note that the apoplast of barley leaves has been shown to contain substantial amounts of superoxide dismutase, ascorbate peroxidase and other enzymes required for destruction of superoxide and H_2O_2 (Vanacker *et al.*, 1998). Antisense ascorbate peroxidase plants showed enhanced ozone sensitivity (Orvar and Ellis, 1997), suggesting that ascorbate peroxidase has a key H_2O_2-scavenging role.

H_2O_2 has been implicated as a key factor mediating the programmed cell death (PCD) that occurs during the HR in plants and also in suspension cultures (Alvarez and Lamb, 1997). Cell death is induced in *Arabidopsis* suspension cultures by harpin and this effect is mediated at least partly by H_2O_2 (Desikan *et al.*, 1996). Exogenous H_2O_2 induces cell death in a dose- and time-dependent manner (*Figure 2a*). That such cell death is programmed is indicated by the requirement for transcription and translation and for a 'presentation time' for exposure to H_2O_2 (Desikan *et al.*, 1998a; *Figure 2b, c*).

As yet, there are little data regarding intracellular signalling processes mediating H_2O_2 responses. Oxidative stress results in increased cytosolic calcium (McAinsh *et al.*, 1996; Price *et al.*, 1994) and H_2O_2-induced PCD in soybean cultures was dependent on calcium influx and protein phosphorylation (see Alvarez and Lamb, 1997). Mitogen-activated protein (MAP) kinases are activated in response to a number of biotic and abiotic stresses (Hirt, 1997). We found that harpin induced the rapid activation of a 44 kDa protein kinase with the characteristics of a MAP kinase and that the same, or a similar, kinase is also activated by H_2O_2 (*Figure 3*). This kinase is activated within 5 min of exposure to H_2O_2 and its activity is decreased again within 30 min (*Figure 3*). It has the characteristics of a MAP kinase in that it phosphorylates myelin basic protein in preference to casein or histone and appears to require tyrosine phosphorylation for activity: it is immunoprecipitated by an anti-phosphotyrosine monoclonal antibody and its activity is substantially reduced following treatment with a phosphotyrosine phosphatase (data not shown). It remains to be established whether or not activation of this protein kinase is essential for H_2O_2-induced PCD and gene expression. However, we have found that

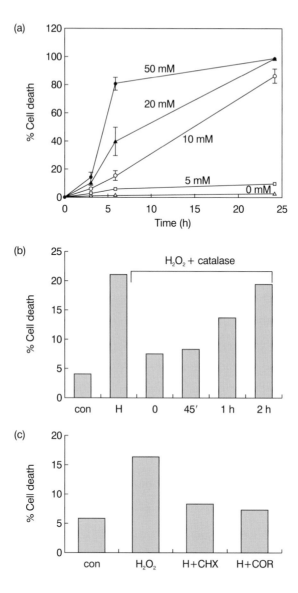

Figure 2. H_2O_2-induced programmed cell death in *Arabidopsis* suspension cultures. (a) Cells were exposed to various concentrations of H_2O_2 and cell death determined at increasing times after exposure. (b) Cells were exposed to 10 mM H_2O_2 (H) and at various times catalase was added to the culture medium. Cell death was then determined 6 h from time zero. (c) Cells were exposed to 10 mM H_2O_2 in the absence (H_2O_2) or presence of cycloheximide (CHX, an inhibitor of translation) or cordycepin (COR, a transcription inhibitor). con, mock-treated cells.

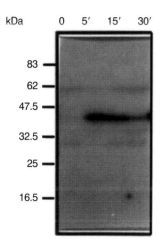

Figure 3. H$_2$O$_2$ induces the activation of a 44 kDa myelin basic protein kinase in *Arabidopsis* suspension cultures. Cells were exposed to 10 mM H$_2$O$_2$ and harvested at increasing times after exposure. Crude protein extracts were then analysed by an in-gel kinase assay using myelin basic protein as the substrate embedded in the gel.

the compound PD98059, which specifically inhibits MAP kinase kinase, and therefore MAP kinase activation in mammalian cells (Cohen, 1997) inhibits both H$_2$O$_2$-induced PCD and gene expression in a dose-dependent manner. Our preliminary data indicate that treatment with PD98059 inhibits the activation of the putative MAP kinase by either harpin or H$_2$O$_2$, but that PD98059 has no effect on its activity.

Current work in our laboratory is directed towards: (i) characterization of the protein kinases activated by harpin and H$_2$O$_2$. What are the sequences of these kinases? What roles do they play in mediation of H$_2$O$_2$-induced gene expression and cell death? The answers to such questions will require purification and sequencing of the kinases and identification of their cellular partners; (ii) identification of the nature of programmed cell death. Does it have features similar to apoptosis as seen in other plant systems? What signalling processes are involved? (iii) What other genes are induced or repressed by H$_2$O$_2$? We are currently analysing a number of genes identified by differential display in the hope that they may shed some light on cellular responses to H$_2$O$_2$.

References

Alvarez, M.E. and Lamb, C. (1997) Oxidative burst-mediated defense responses in plant disease resistance. In: *Oxidative Stress and the Molecular Biology of Antioxidant Defenses* (ed. J.G. Scandalios). Cold Spring Harbor Laboratory Press, Cold Spring Harbor, NY, pp. 815–839.

Bartosz, G. (1997). Oxidative stress in plants. *Acta Physiol. Plant.* **19**: 47–64.

Bestwick, C.S., Brown, I.R., Bennett, M.H.R. and Mansfield, J.W. (1997) Localization of hydrogen peroxide accumulation during the hypersensitive reaction of lettuce cells to *Pseudomonas syringae* pv. *phaseolicola*. *Plant Cell* **9**: 209–221.

Cohen, P. (1997) The search for physiological substrates of MAP and SAP kinases in mammalian cells. *Trends Cell. Biol.* **7**: 353–361.

Dangl, J. (1998) Plants just say NO to pathogens. *Nature* **394**: 525–527.

Desikan, R., Hancock, J.T., Coffey, M.J. and Neill, S.J. (1996) Generation of active oxygen in elicited cells of *Arabidopsis thaliana* is mediated by a NADPH oxidase-like enzyme. *FEBS Lett.* **382**: 213–217.

Desikan, R., Neill, S.J. and Hancock, J.T. (1997) Generation of active oxygen in *Arabidopsis thaliana*. *Phyton* **37**: 65–70.

Desikan, R., Reynolds, A., Hancock, J.T. and Neill, S.J. (1998a) Harpin and hydrogen peroxide both initiate programmed cell death but have differential effects on defence gene expression in *Arabidopsis thaliana* suspension cultures. *Biochem. J.* **330**: 115–120.

Desikan, R., Burnett, E.C., Hancock, J.T. and Neill, S.J. (1998b) Harpin and hydrogen peroxide induce the expression of a homologue of gp91-phox in *Arabidopsis thaliana* suspension cultures. *J. Exp. Bot.* **49**: 1767–1771.

Doke, N. (1997) The oxidative burst: roles in signal transduction and plant stress. In: *Oxidative Stress and the Molecular Biology of Antioxidant Defenses* (ed. J.G. Scandalios). Cold Spring Harbor Laboratory Press, Cold Spring Harbor, NY, pp. 785–813.

Foyer, C.H., Lopez-Delgado, H., Dat, J.F. and Scott, I.M. (1997) Hydrogen peroxide- and glutathione-associated mechanisms of acclimatory stress tolerance and signalling. *Physiol. Plant.* **100**: 241–254.

Hirt, H. (1997). Multiple roles of MAP kinases in plant signal transduction. *Trends Plant Sci.* **2**: 11–15.

Keller, T., Damude, H.G., Werner, D., Doerner, P., Dixon, R.A. and Lamb, C. (1998). A plant homolog of the neutrophil NADPH oxidase gp91phox subunit gene encodes a plasma membrane protein with Ca^{2+} binding motifs. *Plant Cell* **10**: 255–266.

Levine, A., Tenhaken, R., Dixon, R. and Lamb, C. (1994) H_2O_2 from the oxidative burst orchestrates the plant hypersensitive disease resistance response. *Cell* **79**: 583–595.

McAinsh, M.R., Clayton, H., Mansfield, T.A. and Hetherington, A.M. (1996) Changes in stomatal behaviour and guard cell cytosolic free calcium in response to oxidative stress. *Plant Physiol.* **111**: 1031–1042.

Orvar, B.L. and Ellis, B.E. (1997) Transgenic tobacco plants expressing antisense RNA for cytosolic ascorbate peroxidase show increased susceptibility to ozone injury. *Plant J.* **11**: 1297–1305.

Prasad, T.K., Anderson, M.D., Martin, B.A. and Stewart, C.R. (1994) Evidence for chilling-induced oxidative stress in maize seedlings and a regulatory role for hydrogen peroxide. *Plant Cell* **6**: 65–74.

Price, A.H., Taylor, A., Ripley, H.J., Grifiths, A., Trewavas, A.J. and Knight, M.R. (1994) Oxidative signals increase cytosolic calcium. *Plant Cell* **6**: 1301–1310.

Sharma, Y.L., Leon, J., Raskin, I. and Davis, K.R. (1996) Ozone-induced responses in *Arabidopsis thaliana*: the role of salicylic acid in the accumulation of defense-related transcripts and induced resistance. *Proc. Natl Acad. Sci. USA* **93**: 5099–5104.

Smirnoff, N. (1998) Plant resistance to environmental stress. *Curr. Opinion Biotech.* **9**: 214–219.

Torres, M.A., Onouchi, H., Hamada, S., Machida, C., Hammond-Kosack, K.E. and Jones, J.D.G. (1998) Six *Arabidopsis thaliana* homologues of the human respiratory burst oxidase (gp91phox). *Plant J.* **14**: 365–370.

Vallelian-Bindeschedler, L., Schweizer, P., Mosinger, E. and Metraux, J.-P. (1998) Heat-induced resistance in barley to powdery mildew (*Blumeria graminis* f.sp *hordei*) is associated with a burst of active oxygen species. *Physiol. Mol. Plant Pathol.* **52**: 185–199.

Vanacker, H., Carver, T.L.W. and Foyer, C.H. (1998) Pathogen-induced changes in the antioxidant status of the apoplast in barley leaves. *Plant Physiol.* **117**: 1103–1114.

Wojtaszek, P. (1997) Oxidative burst: an early plant response to pathogen infection. *Biochem J.* **322**: 681–692.

Chapter 9

De-repression of heat shock protein synthesis in transgenic plants

Friedrich Schöffl and Ralf Prändl

1. Background

The heat shock (hs) response is a conserved reaction of cells to environmental stresses which is characterized by a rapid induction of the synthesis of hs proteins (HSPs) and acquisition of thermotolerance. In plants a developmentally controlled hs-response occurs also in the absence of hs during seed maturation, pollen formation and microspore embryogenesis.

In *Arabidopsis thaliana* a number of potentially regulatory genes encoding heat shock factors (HSFs) are known. We have used transgenic approaches to investigate and manipulate the expression and regulation of different HSFs with respect to the de-repression of the regulated activity of HSF, the involvement of other regulatory genes, the expression of HSPs and the effect on thermotolerance.

HSPs seem to accumulate in a dosage-dependent manner in response to heat stress and are also induced by a number of other environmental stresses. The major groups of HSPs including HSP100, HSP90, HSP70, HSP60 and small HSP are functionally related. Different HSPs may have different functional properties but common to all of them is their capacity to interact with other proteins and to act as molecular chaperones *in vitro* (for overview see Boston *et al.*, 1996; Schöffl *et al.*, 1998a, 1998b) and probably also *in vivo* as indicated by the protection and reactivation of a luciferase reporter in *Arabidopsis* cells following transient expression of HSPs (Forreiter *et al.*, 1997).

Molecular chaperones are defined by their capacity to recognize and to bind substrate proteins that are kept in an unstable, inactive (denatured) state, competent for activation by proper folding. Chaperone activity is required by all living cells throughout their lifetime. The massive induction upon heat stress indicates a much higher demand, probably resulting from an increase in protein denaturation. Consequently, the cellular chaperone pool has to be replenished after stress.

Expression of small HSPs in the absence of environmental stress is evident during early meiotic stages of pollen development in various species including maize (Atkinson *et al.,* 1993; Magnard *et al.,* 1996), tobacco and tomato (Zur Nieden *et al.,* 1995) and also occurs during embryogenesis from somatic cells, microspores and

Plant Responses to Environmental Stress, edited by M.F. Smallwood, C.M. Calvert and D.J. Bowles.
© 1999 BIOS Scientific Publishers, Oxford.

developing pollen in alfalfa and tobacco (Györgyey *et al.*, 1991; Zársky *et al.*, 1995; for overview see Touraev *et al.*, 1997). Despite largely different conditions of induction, microspore-derived embryos from tobacco and somatic embryos from alfalfa express small HSP during the globular as well as the heart stage but not during the following torpedo stage.

In zygotic embryos, expression of heat shock genes occurs during the maturation stage of the seed when cell division has ceased and seeds adapt to desiccation and long-term survival. It has been proposed that at this stage HSPs are important for desiccation tolerance of the embryo or are required for germination upon rehydration. Similar to other plants, *Arabidopsis* accumulates a specific subset of HSPs during seed maturation including AtHSP17.4, AtHSP17.6 but not AtHSP18.2 (Wehmeyer *et al.*, 1996). The developmental induction of HSP expression in seeds seems to depend on the HSE-containing part of the promoter region of heat shock genes (Prändl and Schöffl, 1996) but also on the activity of ABI3, a regulator of gene expression during seed maturation (Wehmeyer *et al.*, 1996).

The differential expression of heat shock genes during gametogenesis and embryo-genesis compared to heat-induced HSPs suggests that the developmentally expressed HSPs serve certain functions that may differ, at least to some extent, from those required for coping with heat stress in vegetative tissue. These differences in HSP expression between environmentally and developmentally induced heat shock response suggest differences in the signal transduction pathways.

Both, environmentally and developmentally induced expression of HSPs appears to be regulated at the transcriptional level, depending on the activity of one or more HSFs. HSFs are members of a multigene family in plants and can be divided into two sub-groups, A and B, based on the extent of conservation within conserved structural domains (for review see Nover *et al.*, 1996; Schöffl *et al.*, 1998a). The multitude of HSF-like genes expressed in plants is still enigmatic. The major and still unsolved questions are: What are the functional differences of the different HSFs? Which signals are required for environmental and developmental regulation of HSF? What are the molecular mechanisms for de-repression/repression of HSF activity upon heat stress and during development?

2. Structural and functional characteristics of heat shock factors

To date, four HSFs have been described from *A. thaliana* (Hübel and Schöffl, 1994; Nover *et al.*, 1996; Prändl *et al.*, 1998), six from soybean (Czarnecka-Verner *et al.*, 1995), three from tomato (Scharf *et al.*, 1990), three from maize (Gagliardi *et al.*, 1995) and two from tobacco (accession numbers AB014483 and AB014484). The actual number and diversity of expressed HSF-like genes is much higher as exemplified for *Arabidopsis* (*Table 1*). Besides the already described HSF1 (Hübel and Schöffl, 1994), HSF3 and HSF4 (Prändl *et al.*, 1998), a number of additional cDNAs for potential HSFs were isolated and characterized in our laboratory (see *Figure 1*), further genes were identified in cDNA and genomic libraries by conservation of sequences of the DNA-binding domain (see I–VIII, *Table 1*). The conserved features of HSFs, compared with

Table 1. *Arabidopsis* heat shock factors (HSFs) with almost completely sequenced mRNAs (1–7, 21) and putative HSFs identified by partially sequenced cDNAs or by genomic clones (I–VIII)

HSF	GenBank accession numbers[a]		Identity of aa (%)[b]
	cDNA	Genomic DNA	
1		X76167, Z97344	100
2	N96842,[c]		95
3	Y14068		100
4	U68017, Y14069	Z99707	90
5	[c]		85
6	[c]	AB016880	95
7	[c]		100
21	U68561	AL021711	85
I	H37587	AB012245	80
II		AB018115	85
III		AC002510	85
IV		AC004747, B26050	85
V	F15453		90
VI	H36030, T21116		95
VII	T44458		≥80
VIII		Z97335	90

[a]Identified by a blast search in the *Arabidopsis* GenBank Data Set using a conserved HSF sequence (aa 92–111 of HSF1, Hübel and Schöffl, 1994).
[b]Amino acid (aa) identity with the HSF1 sequence (aa 92–111).
[c]Our unpublished data.

the sole HSF of *Drosophila melanogaster* in *Figure 1*, comprise the DNA-binding and the oligomerization domains (hydrophobic regions A and B) located in the N-terminal region of HSF. Both domains are conserved in primary structure throughout the HSF protein family. In other regions significant homology is only detected between closely related HSFs. Nuclear localization signals, a hydrophobic heptad repeat localized in the C-terminal region, and activation domains have been identified by functional studies in several HSFs (for overview see Wu, 1995; Mager and De Krujiff, 1995) including those from tomato (Lyck *et al.*, 1997; Treuter *et al.*, 1993).

The highest degree of conservation is found within the DNA-binding domain consisting of an anti-parallel four-stranded β-sheet and three α-helices (for overview see Wu, 1995; Mager and De Krujiff, 1995; Nover *et al.*, 1996; Schöffl *et al.*, 1998a). The oligomerization domain is characterized by hydrophobic repeat sequences (HR-A/B) separated from the DNA-binding domain by a linker of variable length and sequence. Region HR-A is based on a pattern of seven amino acid repeats of hydrophobic residues, whereas region B is composed of two overlapping stretches of seven amino acid repeats. In class A plant HSFs, these arrays are separated by three seven-amino acid repeats which are missing in class B-HSFs. It is assumed that the hydrophobic repeat A/B-region is required for trimer formation through a triple-stranded, α-helical coiled-coil structure (for overview see Wu, 1995; Mager and De Krujiff, 1995; Nover *et al.*, 1996). The differences in conservation of this domain, discriminating HSFs in classes A and B (*Figure 1*), suggest that functional differences may concern properties

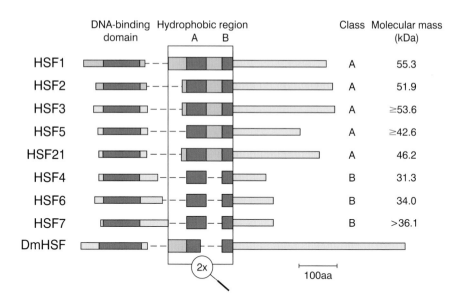

Figure 1. Structural motifs in *Arabidopsis* HSFs. HSF1–HSF21, *Arabidopsis* HSFs; DmHSF, *Drosophila* HSF.

of multimerization, complex formation and consequently also DNA-binding. In higher eukaryotes, including plants, the activation of HSF for binding to the conserved hs promoter elements (HSE) with the consensus sequence $(-GAA–TTC-)_n$, depends on the heat-inducible formation of HSF trimers. The unifying features of HSF1 in different organisms are its constitutive expression and heat-dependent activation involving trimerization, DNA-binding and transcriptional activation of hs genes.

Another intriguing difference between class A- and class B-HSFs of *Arabidopsis* (*Figure 1*) is in the calculated molecular masses ranging in class A from approximately 45 to 55 kDa and in class B from 30 to 35 kDa. The smaller size of class B-HSFs results from shortage of sequences within the less conserved C-terminal part of HSFs. The C-terminal half of HSF1, for example, harbours the *trans*-activation domain and sequences involved in regulating repression/de-repression of HSF conformation and activity (Green *et al.*, 1995; Newton *et al.*, 1996; Shi *et al.*, 1995; Wisniewski *et al.*, 1996). The size differences between class A- and class B-HSFs may be not as strict in other plant species (for comparison see Nover *et al.*, 1996), however, structural and functional properties and differences have still to be determined. To date there is no evidence that HSFs belonging to class B in plants are transcriptional activators.

It has been proposed by Czarnecka-Verner *et al.* (1997) that the positive (activation of transcription) and negative (repression of HSF activity) aspects of HSF regulation, intrinsic to animal and human HSF, are separated in plants and specified by individual HSF proteins of class A (activator HSFs) and class B (inert or repressor HSFs). According to this model, constitutive class B-HSFs could act as repressors at normal temperature and heat-inducible B-HSFs might be involved in de-activation of hs genes. It has been shown that the nuclear transport of a heat inducibly expressed tomato

HSF-A2 is dependent on an interaction with the constitutive HSF-A1, the B-type HSF-B1 was unable to substitute HSF-A1 in this function (Scharf *et al.*, 1998). HSFs carry two clusters of basic amino acids that have been proposed to function as nuclear localization sequences. In tobacco, recombinant tomato HSFs required only the more C-terminal nuclear localization sequence for nuclear import (Lyck *et al.*, 1997).

In animal and human cells it is known that a C-terminally localized hydrophobic repeat (not depicted in *Figure 1*) is involved in the regulation of trimerization of HSF1 and HSF3 (Nakai and Morimoto, 1993; Rabindran *et al.*, 1993; Zuo *et al.*, 1994). This region is poorly conserved in plant and yeast HSFs and it is not known which sequences in this region play an important role in the regulation of HSF conformation and multimerization in these organisms.

3. Regulation of heat shock factor activity

There is evidence for trans-acting regulators of HSF activity in different organisms, including plants. The transient nature of the hs response, constitutive HSP synthesis in mutants in HSP70/HSC70 genes in yeast (Boorstein and Craig, 1990), prolonged activity of HSF in HSP70 antisense plants (Lee and Schöffl, 1996), interactions between HSF and HSP70 in mammalian cells (Baler *et al.*, 1992; Shi *et al.*, 1998) and plants B.-H. Kim and F. Schöffl, unpublished data) and HSP90 in *Xenopus* oocytes (Ali *et al.*,

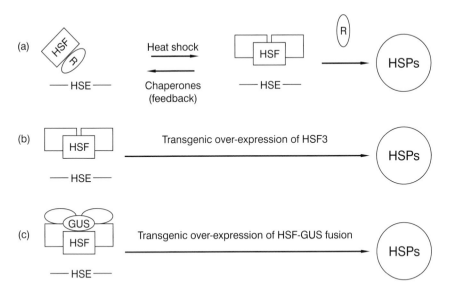

Figure 2. Model of mechanism of de-repression of HSF activity. (a) Feedback regulation: repression of HSF trimerization and DNA-binding by a transacting negative regulator (chaperone R) at normal temperature; de-repression upon heat stress via dissociation of R from HSF. (b) Chaperone titration by excess HSF (upon transgenic over-expression of HSF) results in trimeric DNA-binding and active HSF at normal temperature. (c) Same as in (b) resulting from transgenic expression of HSF-GUS fusion proteins. HSF, heat shock transcription factor; R, 'repressor' chaperone; GUS, glucuronidase reporter; HSP, heat shock proteins; HSE, heat shock promoter element (HSF binding sequence).

1998) are consistent with a negative feedback control of HSF activity (*Figure 2a*), probably in a complex activation/de-activation pathway (Nunes and Calderwood, 1995).

Covalent modifications of HSF, for example phosphorylation, has been proposed to play a role in activation and inactivation of HSFs (for overview see Wu, 1995; Mager and Krujiff, 1995). Phosphorylation of serine residues is increased upon stimulation of the Raf/ERK pathway, a mitogen-activated protein kinase pathway responsive to growth factors, and results in inhibition of HSF1 activity in mammalian cells (Knauf *et al.*, 1996; Chu *et al.*, 1996). In plants, serine phosphorylation of recombinant AtHSF1 by a CDC2a cyclin-dependent kinase has been demonstrated in extracts of *Arabidopsis* suspension culture (Reindl *et al.*, 1997). *In vitro* phosphorylated HSF fractions showed a decreased capacity of HSE-binding. Thus, in human cells as well as in *Arabidopsis*, phosphorylation of HSF through various kinases may integrate growth signals, and it is conceivable that in growing cells phosphorylation of HSF is required for repression of the hs response which might otherwise interfere with cell proliferation.

4. Transgenic expression and de-repression of heat shock factors in plants

HSF1 is repressed under non-stress conditions and trimerizes upon heat shock. Upon expression in transgenic plants two different but related modes of de-repression (activation) of HSF activity were detected (*Figure 2b and* c). The criteria for activation were: DNA-binding to HSE sequences, transcription of hs genes, synthesis of HSPs at non-hs temperatures and increased basal thermotolerance.

A heat stress-independent de-repression of *Arabidopsis* HSFs was obtained by constitutive over-expression of HSF-β-glucuronidase (HSF-GUS) fusion proteins using AtHSF1 (Lee *et al.*, 1995) or AtHSF3 (Prändl *et al.*, 1998); both HSFs belong to subgroup A. The exact molecular mechanism of de-repression is still unknown but seems not only restricted to GUS fusions of HSF (S. Döhr and F. Schöffl, unpublished data). In fusion proteins the conformation of HSF may be altered and consequently becomes inaccessible to a negative regulatory factor. Interestingly, ectopic over-expression of AtHSF3 in transgenic *Arabidopsis* was sufficient to de-repress HSF activity (Prändl *et al.*, 1998). Over-expression of HSF may not result in a conformational alteration and HSF activation directly, it seems possible that de-repression is a consequence of titrating a *trans*-acting negative regulator by excess HSF. *Arabidopsis* HSF1 shows also constitutive DNA-binding upon heterologous expression in *Drosophila* and human cells and was able to activate transcription of a suitable reporter gene in *Drosophila* (Hübel *et al.*, 1995). Thus, the negative control of HSF in homologous plant cells seems to depend on a specific, as yet unknown factor that is obviously absent in the cultured animal cells. HSP70 and HSP90 chaperones and hydrophobic repeat HSF-binding proteins such as, for example, the animal HSBP1 (Satyal *et al.*, 1998), seem to be involved in the feedback regulation, de-activation and recycling of HSF activity (see above), but other more specific factors or modifications may be important.

On the other hand, over-expression of AtHSF4 (a class B-HSF) or AtHSF4-GUS fusion proteins was not sufficient to de-repress the synthesis of HSPs at normal temperature in transgenic *Arabidopsis* (Prändl *et al.*, 1998). This lack of evidence for transcription activating competence of class B-HSFs is in accordance with the negative result obtained for transiently expressed soybean class B-HSFs in tobacco (Czarnecka-Verner *et al.*, 1997).

5. Concluding remarks

The regulation of HSF activity and the multiplicity and biological role of the different HSFs in plants are still not understood and of continuing scientific interest. The genetic redundancy of HSF reflects diversification and it seems possible that some HSFs, particularly those of class B, classified by the criterion of structural features in the DNA-binding and multimerization domains, may function as repressors of transcription of hs genes and/or trans-regulators of the real activator HSF activity.

Another question that remains open is whether there is a stress-sensing signal pathway that triggers the activation of HSFs. Components in the pathway upstream from HSF are not yet known. It is conceivable that HSF itself or in a complex with HSC70, HSP90 and other proteins is the cellular thermometer. Heat stress, developmental signals and recombinant over-expression result in an activation of HSF via conformational changes involving monomer to trimer transition, nuclear targeting, DNA-binding and transcriptional activation. Developmental signalling seems to be responsible for the expression of HSPs during seed maturation. The involvement of HSF is indicated by the dependence of HSE promoter sequences in developmental de-repression of the hs response, and signalling through ABA pathways is suggested by the negative effect of an *abi3* mutation in *Arabidopsis*. Future research will focus on the identification of the responsible HSF and the level of control by ABI3. There are preliminary data suggesting involvement of both ABI3 and HSF in DNA-binding in seeds (unpublished data).

Another pathway may exist that integrates signals of cell proliferation and results in cell cycle-dependent phosphorylation of HSF via Cdc2a (Reindl *et al.*, 1997). It will be important to find out whether HSF phosphorylation also occurs *in vivo* and whether it has an important biological function in cell growth and development.

The transgenic approaches leading to the de-repression of the hs response, independent from environmental stresses, provide evidence that the HSPs are determinants of thermotolerance. Up to now most cellular targets of HSPs are unknown. The current data suggest that HSPs protect other proteins and cellular structures from irreversible denaturation and cytotoxic effects. However, other as yet unknown HSF-dependent up- or down-regulated genes may contribute to this phenotype. Further analysis of genes/proteins which, besides HSPs, are differentially expressed becomes accessible in HSF de-repressed transgenic lines.

Acknowledgements

The research work was supported by grants of the Deutsche Forschungsgemeinschaft to F.S.

References

Ali, A., Bharadwaj, S., O'Carrol, R. and Ovsenek, N. (1998) HSP90 interacts with and regulates the activity of heat shock factor 1 in *Xenopus* oocytes. *Mol. Cell Biol.* **18**: 4949–4960.

Atkinson, B.G., Raizada, M., Bouchard, R.A., Frappier, R.A. and Walden, D.B. (1993) The independent stage-specific expression of the 18 kDa heat shock protein genes during microsporogenesis in *Zea mays* L. *Dev. Genet.* **14**: 15–26.

Baler, R., Welch, W.J. and Veollmy, R. (1992) Heat shock gene regulation by nascent polypeptides and denatured proteins: hsp70 as a potential autoregulatory factor. *J. Cell Biol.* **117**: 1151–1159.

Boorstein, W.R. and Craig, E.A. (1990) Transcriptional regulation of SSA3, a HSP70 gene from *Saccharomyces cerevisiae. Mol. Cell. Biol.* **10**: 3262–3267.

Boston, R.S., Viitanen, P.V. and Vierling, E. (1996) Molecular chaperones and protein folding in plants. *Plant Mol. Biol.* **32**: 191–222.

Chu, B., Soncin, F., Price, B.D., Stevenson, M.A. and Calderwood, S.T. (1996) Sequential phosphorylation by mitogen-activated protein kinase and glycogen synthase kinase 3 represses transcriptional activation by heat shock factor 1. *J. Biol. Chem.* **271**: 30847–30857.

Czarnecka-Verner, E., Yuan, C.X., Fox, P.C. and Gurley, W.B. (1995) Isolation and characterization of six heat shock transcription factor cDNA clones from soybean. *Plant. Mol. Biol.* **29**: 37–51.

Czarnecka-Verner, E., Yuan, C.X., Nover, L., Scharf, K.D., English, G. and Gurley, W.B. (1997) Plant heat shock transcription factors: positive and negative aspects of regulation. *Acta Physiol. Plant.* **19**: 529–537.

Forreiter, C., Kirschner, M. and Nover, L. (1997) Stable transformation of an Arabidopsis cell suspension culture with firefly luciferase providing a cellular system for analysis of chaperone activity *in vivo. Plant Cell* **9**: 2171–2181.

Gagliardi, D., Breton, C., Chaboud, A., Vergne, P. and Dumas, C. (1995) Expression of heat shock factor and heat shock protein 70 genes during maize pollen development. *Plant Mol. Biol.* **29**: 841–856.

Green, M., Schuetz, T.J., Sullivan, E.K. and Kingston, R.B. (1995) A heat shock-reponsive domain of human HSF1 that regulates transcription activation domain function. *Mol. Cell Biol.* **15**: 3354–3362.

Györgyey, J., Gartner, A., Németh, K., Magyar, Z., Hirt, H., Heberle-Bors, E. and Dudits, D. (1991) Alfalfa heat shock genes are differentially expressed during somatic embryogenesis. *Plant Mol. Biol.* **16**: 999–1007.

Hübel, A. and Schöffl, F. (1994) Arabidopsis heat shock factor: characterization of the gene and the recombinant protein. *Plant Mol. Biol.* **26**: 353–362.

Hübel, A., Lee, J.H., Wu, C. and Schöffl, F. (1995) *Arabidopsis* heat shock factor is constitutively active in *Drosophila* and human cells. *Mol. Gen. Genet.* **248**: 136–141.

Knauf, U., Newton, E.M., Kyriakis, J. and Kingston, R.E. (1996) Repression of human heat shock factor 1 activity at control temperature by phosphorylation. *Genes Dev.* **10**: 2782–2793.

Lee, J.H., Hübel, A. and Schöffl, F. (1995) Derepression of the activity of genetically engineered heat shock factor causes constitutive synthesis of heat shock proteins and increased thermotolerance in transgenic *Arabidopsis. Plant J.* **8**: 603–612.

Lee, J.H. and Schöffl, F. (1996) An *Hsp70* antisense gene affects the expression of HSP70/HSC70, the regulation of HSF, and the acquisition of thermotolerance in transgenic *Arabidopsis thaliana. Mol. Gen. Genet.* **252**: 11–19.

Lyck, R., Harmening, U., Höhfeld, I., Treuter, E., Scharf, K.D. and Nover, L. (1997) Intracellular distribution and identification of the nuclear localization signals of two plant heat-stress transcription factors. *Planta* **202**: 117–125.

Mager, W.H. and De Krujiff, A.J.J. (1995) Stress-induced transcriptional activation. *Microbiol. Rev.* **59**: 506–532.

Magnard, J.L., Vergne, P. and Dumas, C. (1996) Complexity and genetic variability of heat shock protein expression in isolated maize microspores. *Plant Physiol* **111**: 1085–1096.

Nakai, A. and Morimoto, R.I. (1993) Characterization of a novel chicken heat shock transcription factor, heat shock factor 3, suggests a new regulatory pathway. *Mol. Cell Biol.* **13**: 1983–1997.

Newton, E.M., Knauf, U., Green, M. and Kingston, R.E. (1996) The regulatory domain of human heat shock factor 1 is sufficient to sense heat stress. *Mol. Cell Biol.* **16:** 839–846.

Nover, L., Scharf, K.D., Gagliardi, D., Vergne, P., Czarnecka-Verner, E. and Gurley, W.B. (1996) The Hsf world: classification of plant heat stress transcription factors. *Cell Stress* **1:** 215–223.

Nunes, S.L. and Calderwood, S.K. (1995) Heat shock factor-1 and the heat shock cognate 70 protein associate in high molecular weight complexes in the cytoplasm of NIH-3T3 cells. *Biochem. Biophys. Res. Comm.* 213: 1–6.

Prändl, R. and Schöffl, F. (1996) Heat shock elements are involved in heat shock promoter activation during tobacco seed maturation. *Plant Mol. Biol.* **31:** 157–162.

Prändl, R., Hinderhofer, K., Eggers-Schumacher, G. and Schöffl, F. (1998) HSF3, a new heat shock factor from *Arabidopsis thaliana* derepresses the heat shock response and confers thermotolerance in transgenic plants. *Mol. Gen. Genet.* **258:** 269–278.

Rabindran, S.K., Haroun, R.I., Clos, J., Wisniewski, J. and Wu, C. (1993) Regulation of heat shock factor trimer formation: role of a conserved leucine zipper. *Science* **259:** 230–234.

Reindl, A., Schöffl, F., Schell, J., Koncz, C. and Bako, L. (1997) Phosphorylation by a cyclin-dependent kinase modulates DNA-binding of the *Arabidopsis* heat shock transcription factor HSF1 *in vitro*. *Plant Physiol.* **115:** 93–100.

Satyal, S.H., Chen, D., Fox, S.G., Kramer, J.M. and Morimoto, R.I. (1998) Negative regulation of the heat shock transcriptional response by HSBP1. *Genes Dev.* **12:** 1962–1974.

Scharf, K.D., Rose, S., Zott, W., Schöffl, F. and Nover, L. (1990) Three tomato genes code for heat stress transcription factors with a region of remarkable homology to the DNA-binding domain of yeast HSF. *EMBO J.* **9:** 4495–4501.

Scharf, K.D., Heider, H., Höfeld, I., Lyck, R., Schmidt, E. and Nover, L. (1998) The tomato Hsf system: HsfA2 needs interaction with HsfA1 for efficient nuclear import and may be localized in cytoplasmic heat stress granules. *Mol. Cell Biol.* **18:** 2240–2251.

Schöffl, F., Hübel, A. and Lee, J.H. (1998a) Manipulation of temperature stress tolerance in transgenic plants. In: *Transgenic Plant Research* (ed. K. Lindsey). Harwood Academic Publishers, London, pp. 229–240.

Schöffl, F., Prändl, R. and Reindl, A. (1998b) Regulation of the heat-shock response. *Plant Physiol.* **117:** 1135–1141.

Shi, Y.H., Kroeger, P.E. and Morimoto, R.I. (1995) The carboxy-terminal transactivation domain of heat shock factor 1 is negatively regulated and stress responsive. *Mol. Cell Biol.* **15:** 4309–4318.

Shi, Y., Mosser, D.D. and Morimoto, R.I. (1998) Molecular chaperones as HSF1-specific transcriptional repressors. *Genes Dev.* **12:** 654–666.

Touraev, A., Vicente, O. and Heberle-Bors, E. (1997) Initiation of microspore embryogenesis by stress. *Trends Plant Sci.* **2:** 297–302.

Treuter, E., Nover, L., Ohme, K. and Scharf, K.D. (1993) Promoter specificity and deletion analysis of three heat shock transcription factors of tomato. *Mol. Gen. Genet.* **240:** 113–125.

Wehmeyer, N., Hernandez, L.D., Finkelstein, R.R. and Vierling, E. (1996) Synthesis of small heat shock proteins is part of the developmental program of late seed maturation. *Plant Physiol.* **112:** 747–757.

Wisniewski, J., Orosz, A., Allada, R. and Wu, C. (1996) The C-terminal region of *Drosophila* heat shock factor (HSF) contains a constitutively functional transactivation domain. *Nucleic Acids Res.* **24:** 367–374.

Wu, C. (1995) Heat schock transcription factors: structure and regulation. *Annu. Rev. Cell Dev. Biol.* **11:** 441–469.

Zársky, V., Garrido, D., Eller, N., Tupy, J., Vicente, O., Schöffl, F. and Heberle-Bors, E. (1995) The expression of a small heat shock gene is activated during induction of tobacco pollen embryogenesis by starvation. *Plant Cell Environ.* **18:** 139–147.

Zuo, J., Baler, R., Dahl, G. and Voellmy, R. (1994) Activation of the DNA-binding ability of human heat shock transcription factor 1 may involve the transition from an intramolecular to an intermolecular triple-stranded coiled-coil structure. *Mol. Cell Biol.* **14:** 7557–7568.

Zur Nieden, U., Neumann, D., Bucka, A. and Nover, L. (1995) Tissue specific localization of stress proteins during embryo development. *Planta* **196:** 530–538.

Chapter 10

Regulation of plant cold acclimation

Michael F. Thomashow

1. Background

During the warm growing seasons, plants are generally very sensitive to freezing. However, as the year progresses, many plants in temperate regions sense changes in the environment that signal the coming winter and respond by increasing in freezing tolerance. The primary environmental factor responsible for this response is low non-freezing temperatures; the response is known as cold acclimation.

How do plants sense low temperature and activate the cold acclimation response? Determining the answer to this question will not only add to our fundamental understanding of how plants adapt to changes in the environment, but also has potential practical applications. Freezing temperatures are a major factor determining the geographical locations suitable for growing crop and horticultural plants and periodically account for significant losses in productivity. Determining the mechanisms whereby plants sense low temperature and activate the cold acclimation response has the potential to provide new strategies for improving the freezing tolerance of agronomic plants. Such strategies would be highly significant as traditional plant breeding approaches have had limited success in improving freezing tolerance. For instance, despite considerable effort, the freezing tolerance of wheat varieties today is only marginally better than those developed in the early part of this century (Fowler and Gusta, 1979).

Classical genetic analyses indicated early on that the ability of plants to cold acclimate is a quantitative trait largely involving the action of many genes, each with small additive effects (Thomashow, 1990). Until recently, however, there has been little progress in identifying specific genes with roles in freezing tolerance. Classical genetic approaches have been hampered by multiple factors including the quantitative nature of the trait as well as that an accurate determination of freezing tolerance requires the testing of plant populations. However, by using newly developed molecular marker technologies and methods of QTL (quantitative trait locus) mapping, exciting new information on freezing tolerance loci has begun to emerge, particularly in wheat (Galiba *et al.*, 1995) and barley (Hayes *et al.*, 1993). In each of these species, a locus has been identified that has a major effect on freezing tolerance. Moreover, these loci appear to be related. The locus, known as the Vrn1-Fr1 interval in wheat, may encode a

Plant Responses to Environmental Stress, edited by M.F. Smallwood, C.M. Calvert and D.J. Bowles.

regulatory gene(s) that controls the level of expression of cold-inducible genes (Limin *et al.*, 1997). The challenge now is to determine the identity and functions of the genes at these freezing tolerance loci. This will not be an easy task, but the expanding efforts in cereal genomic research should greatly facilitate these efforts.

In 1985, Guy *et al.* established that changes in gene expression occur during cold acclimation. This quickly led to a new line of investigation, the isolation and characterization of genes that are induced during cold acclimation. The notion was that some genes that are activated in response to low temperature probably have roles in metabolic adjustment to low temperatures, but that others might have specific roles in freezing tolerance. Indeed, recent studies with *Arabidopsis* have led to the identification of a freezing tolerance gene 'regulon' and a family of transcription factor genes that control their expression. This work is summarized in detail below. In addition, McKown *et al.* (1996) and Xin and Browse (1998) recently initiated another powerful line of investigation, the isolation of *Arabidopsis* mutants altered in freezing tolerance. These studies have already led to the identification of a number of loci that have dramatic effects on freezing tolerance. Given the resources available for *Arabidopsis* – EST (expressed sequence tag) libraries; BAC, YAC and cosmid genomic libraries with defined contig groups; rapidly increasing genomic sequence information; molecular maps; T-DNA and transposon insertion mutagenized lines and so on – rapid progress should be made in identifying the nature of the freezing tolerance genes.

2. Identification and regulation of *Arabidopsis* freezing tolerance genes

As alluded to above, one approach being taken to identify potential freezing tolerance genes is to isolate and characterize genes that are induced in response to low temperature. This approach has recently resulted in the identification of a freezing tolerance gene 'regulon' in *Arabidopsis* and a family of transcription factor genes that control their expression.

2.1 Role of **Arabidopsis** *COR genes in freezing tolerance*

Among the highly expressed cold-responsive genes of *Arabidopsis* are the *COR* (cold-regulated) genes, also designated *LTI* (low temperature-induced), *KIN* (cold-inducible), *RD* (responsive to desiccation) and *ERD* (early dehydration-inducible) (see Thomashow, 1998). The *COR* genes comprise four gene families, each of which is composed of two genes that are physically linked in the genome in tandem array. The *COR78*, *COR15* and *COR6.6* gene pairs encode newly discovered polypeptides while the *COR47* gene pair encodes LEA (late embryogenesis abundant) group II proteins, also known as LEA D11 proteins, RAB proteins and dehydrins (Ingram and Bartels, 1996).

Artus *et al.* (1996) have provided direct evidence that the *COR15a* gene has a role in cold acclimation. *COR15a* encodes a 15 kDa polypeptide, COR15a, that is targeted to the stromal compartment of chloroplasts (Lin and Thomashow, 1992; S.J. Gilmour and M.F. Thomashow, unpublished data). During import, COR15a is processed to a 9.4 kDa polypeptide, designated COR15am. The COR15am polypeptide is hydrophilic, acidic (pI 4.5), rich in alanine, lysine, glutamic acid and aspartic acid residues, and is predicted to form an amphipathic α-helix. To determine whether *COR15a* might be a freezing-tolerance gene, Artus *et al.* (1996) made transgenic *Arabidopsis* plants that constitutively express *COR15a* and compared the *in vivo* freezing tolerance of the chloroplasts in non-acclimated transgenic and wild-type plants (the level of COR15am in the non-acclimated transgenic plants approximated that in cold-acclimated wild-type plants). Chloroplast freezing tolerance was assessed by freezing plant leaves at various subzero temperatures, and after thawing, estimating the damage to photosystem II by measuring room temperature chlorophyll fluorescence (F_v/F_m). The results indicated that the chloroplasts of the non-acclimated transgenic plants were 1–2°C more freezing tolerant than the chloroplasts from the non-acclimated wild-type plants. In addition, they found that the effects of COR15am were not limited to the chloroplasts. In particular, protoplasts isolated from leaves of non-acclimated transgenic and wild-type plants were frozen *in vitro* at various subzero temperatures, thawed, and the percentage survival was determined by staining with fluorescein diacetate. The results indicated that over the temperature range of –5 to –8°C, the freezing tolerance of the protoplasts from transgenic plants was about 1°C greater than that of the protoplasts prepared from wild-type plants.

By what mechanism does *COR15a* increase freezing tolerance? The results of Artus *et al.* (1996) indicated that constitutive *COR15a* expression increases the cryostability of the plasma membrane. This followed from the fact that protoplast survival was measured by fluorescein diacetate staining, a method that reports on retention of the semipermeable characteristics of the plasma membrane. It was suggested that *COR15a* expression might decrease the propensity of membranes to form hexagonal II phase lipids. Indeed, Steponkus *et al.* (1998) have recently shown that constitutive expression of *COR15a* decreases the incidence of freeze-induced lamellar-to-hexagonal II phase transitions that occur in regions where the plasma membrane is brought in close apposition with the chloroplast envelop as a result of freeze-induced dehydration. In addition, they showed that purified preparations of COR15am increase the lamellar-to-hexagonal II phase transition temperature of dioleoylphosphatidylethanolamine and promotes formation of the lamellar phase in a lipid mixture composed of the major lipid species that comprise the chloroplast envelop. Steponkus *et al.* (1998) have proposed that the COR15am polypeptide acts to defer freeze-induced formation of the hexagonal II phase to lower temperatures by altering the intrinsic curvature of the inner membrane of the chloroplast envelope.

Do any other *Arabidopsis COR* genes have roles in freezing tolerance? To address this question, Jaglo-Ottosen *et al.* (1998) created transgenic *Arabidopsis* plants that constitutively express the entire battery of *COR* genes and compared the freezing tolerance of these plants to those that express *COR15a* alone. Expression of the *COR* gene regulon was accomplished by over-expressing the *Arabidopsis* transcriptional activator CBF1 (<u>C</u>RT/DRE <u>b</u>inding <u>f</u>actor <u>1</u>) (Stockinger *et al.*, 1997). This factor, discussed in detail below, binds to the CRT (<u>C</u>-<u>repeat</u>)/DRE (drought-responsive element) DNA regulatory element present in the promoters of the *COR* genes and activates their expression without a low temperature stimulus. Using an electrolyte leakage test to assess the freezing tolerance of detached leaves from non-acclimated plants, Jaglo-Ottosen *et al.* (1998) were unable to detect a statistically significant enhancement of freezing tolerance by expressing *COR15a* alone. In sharp contrast, they detected a 3°C increase in freezing tolerance in plants that over-expressed CBF1 and consequently, the CRT/DRE *COR* gene regulon. In addition, expression of the CRT/DRE *COR* gene regulon resulted in an increase in whole plant freezing survival whereas expression of *COR15a* alone did not. Taken together, these results implicate additional *COR* genes in freezing tolerance.

Liu *et al.* (1998) have independently shown that expression of the CRT/DRE-regulated *COR* genes increases the freezing tolerance of *Arabidopsis* plants. In their case, they activated *COR* gene expression by over-expressing a homologue of *CBF1* designated *DREB1A*. Significantly, the results of Liu *et al.* (1998) indicate that expression of the CRT/DRE-containing *COR* genes not only affects freezing tolerance, but also increases tolerance to drought. One important difference between the results of Jaglo-Ottosen *et al.* (1998) and Liu *et al.* (1998) is that the latter investigators found that over-expression of *DREB1A* resulted in the transgenic plants having a dwarf phenotype. This was not observed by Jaglo-Ottosen *et al.* (1998) in plants that over-expressed *CBF1*. The reason for this difference is not yet known.

2.2 Regulation of Arabidopsis COR genes

Gene fusion studies have demonstrated that the promoters of the *Arabidopsis COR15*a, *COR6.6* and *COR78* genes are induced in response to low temperature. The cold regulatory element that appears to be primarily responsible for this regulation was first identified by Yamaguchi-Shinozaki and Shinozaki (1994) in their study of the *COR78* promoter. It is a 9 bp element referred to as the DRE. The DRE, which has a 5 bp core sequence of CCGAC designated the CRT (Baker *et al.,* 1994), stimulates gene expression in response to low temperature, drought and high salinity, but not in response to exogenous application of abscisic acid (Yamaguchi-Shinozaki and Shinozaki, 1994). The element, which is also referred to as the LTRE (low temperature regulatory element), has been shown to impart cold-regulated gene expression in *Brassica napus* (Jiang *et al.,* 1996).

Stockinger *et al.* (1997) isolated the first cDNA for a protein that binds to the CRT/DRE sequence. The protein, CBF1, has a mass of 24 kDa, an acidic region that potentially serves as an activation domain and a putative bipartite nuclear localization sequence. In addition, it has an AP2 domain, a 60 amino acid motif that has been found in a large number of plant proteins including *Arabidopsis* APETALA2, AINTEGUMENTA, and TINY; the tobacco EREBPs (ethylene response element binding proteins); and numerous other plant proteins of unknown function (see Riechmann and Meyerowitz, 1998). Ohme-Takagi and Shinshi (1995) have demonstrated that the AP2 domain includes a DNA-binding region.

Stockinger *et al.* (1997) established that the CBF1 protein binds to the CRT/DRE sequence and activates expression of reporter genes in yeast carrying the CRT/DRE as an upstream regulatory sequence. These results indicated that CBF1 is a transcriptional activator that can activate CRT/DRE-containing genes and thus, was a probable regulator of *COR* gene expression in *Arabidopsis*. Subsequently, Jaglo-Ottosen *et al.* (1998) showed that constitutive over-expression of CBF1 in transgenic *Arabidopsis* plants results in expression of the *COR* genes without a low temperature stimulus. Thus, CBF1 appears to be an important regulator of the cold acclimation response, controlling the level of *COR* gene expression which in turn promotes freezing tolerance.

The results of Stockinger *et al.* (1997) and Jaglo-Ottosen *et al.* (1998) have recently been extended by Gilmour *et al.* (1998) and Liu *et al.* (1998). These investigators have established that *CBF1* is a member of a small gene family comprised of three closely related transcriptional activators referred to as CBF1, CBF2 and CBF3 by Gilmour *et al.* (1998) and DREB1B, DREB1C and DREB1A, respectively, by Liu *et al.* (1998). The three *CBF* genes are physically linked in direct repeat on chromosome 4 near molecular markers PG11 and m600 (72.8 cM) (Gilmour *et al.*, 1998). They are unlinked to their target genes *COR6.6*, *COR15a*, *COR47* and *COR78* which are located on chromosomes 5, 2, 1 and 5, respectively (Gilmour *et al.*, 1998). Like CBF1, both CBF2 and CBF3 can activate expression of reporter genes in yeast that contain the CRT/DRE as an upstream activator sequence indicating that these two family members are also transcriptional activators. Moreover, Liu *et al.* (1998) have shown that over-expression of DREB1A (CBF3) in transgenic *Arabidopsis* plants results in constitutive expression of *COR78* and, as described above, enhances both the freezing and drought tolerance of the transgenic plants.

Gene expression studies have revealed that the transcript levels for all three *CBF/DREB1* genes increase dramatically within 15 min of transferring plants to low temperature followed by accumulation of *COR* gene transcripts at about 2 h (Gilmour *et al.*, 1998; Lin *et al.*, 1998). Thus, Gilmour *et al.* (1998) have suggested that *COR* gene expression involves a low temperature signalling cascade in which *CBF* gene expression is an early step. Regulation of the *CBF/DREB1* genes appears to occur, at least in part, at the transcriptional level as hybrid genes containing the *CBF/DREB1* promoters fused to reporter genes are induced at low temperature (D. Zarka and M. Thomashow, unpublished data). As noted by Gilmour *et al.* (1998), the fact that *CBF* transcripts begin accumulating within 15 min of plants being exposed to low temperature strongly suggests that the low temperature 'thermometer' and 'signal transducer' are present at warm non-inducing temperatures. Thus, Gilmour *et al.* (1998) posit that there is a transcription factor already present at warm temperature that recognizes the *CBF* promoters. This factor would not appear to be the CBF proteins themselves as the promoters of the *CBF* genes lack the CRT/DRE sequence and that over-expression of *CBF1* does not cause accumulation of *CBF3* transcripts (Gilmour *et al.*, 1998). Gilmour *et al.* (1998) have, therefore, proposed that *COR* gene induction involves a two-step cascade of transcriptional activators in which the first step, CBF induction, involves an unknown activator that they tentatively designated 'ICE' (inducer of CBF expression) (*Figure 1*). ICE presumably recognizes a cold regulatory element present in the promoters of each *CBF* gene. At warm temperatures, ICE is suggested to be in an 'inactive' state, either because it is sequestered in the cytoplasm by a negative regulatory protein or is in a form which does not bind to DNA or activate transcription effectively. Upon exposing a plant to low temperature, however, a signal transduction pathway is proposed to be activated which

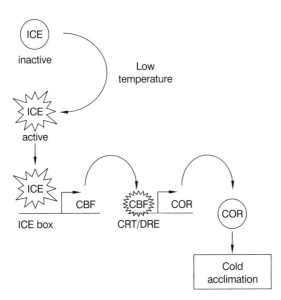

Figure 1. Model for *COR* gene induction in response to low temperature. See text for details. Modified from Gilmour, S.J. *et al.* (1998) Low temperature regulation of the *Arabidopsis* CBF family of AP2 transcriptional activators as an early step in cold-induced CoR gene expression. *Plant Journal*, vol. 16, pp. 433–442. Reprinted with permission of Blackwell Science Ltd.

results in modification of either ICE or an associated protein which in turn allows ICE to induce *CBF* gene expression. The CBF proteins would then be synthesized, bind to the CRT/DRE elements in the promoters of the *COR* (and presumably other) genes, and activate their expression.

3.   Concluding remarks

Recent studies indicate that cold acclimation in *Arabidopsis* involves the induction of the CRT/DRE gene regulon which is brought about by low temperature induction of the CBF family of transcriptional activators. A major goal now is to determine whether freezing tolerance genes in other plants that cold acclimate are organized into CRT/DRE-containing regulons that are controlled by orthologues of the *CBF* regulatory genes. If so, there is the possibility that *CBF* genes could be used as 'master switches' to control the expression of freezing tolerance genes and thereby potentially increase the freezing tolerance of agronomic plants. Other important goals include determining whether genes in the CRT/DRE regulon account for most or only a portion of those with critical roles in cold acclimation and determining how *CBF* gene expression is activated by low temperature. The answers to these questions will add greatly to our understanding of how plants regulate the cold acclimation response.

Acknowledgements

I thank Eric Stockinger and Dan Zarka for suggestions on how to improve the manuscript. Research conducted in the author's laboratory was funded in part by grants from the USDA, NSF and Michigan Agricultural Research Station.

References

Artus, N.N., Uemura, M., Steponkus, P.L., Gilmour, S.J., Lin, C.T. and Thomashow, M.F. (1996) Constitutive expression of the cold-regulated *Arabidopsis thaliana COR15a* gene affects both chloroplast and protoplast freezing tolerance. *Proc. Natl Acad. Sci. USA* **93**: 13404–13409.

Baker, S.S., Wilhelm, K.S. and Thomashow, M.F. (1994) The 5′-region of *Arabidopsis thaliana cor15a* has *cis*-acting elements that confer cold-, drought- and ABA-regulated gene expression. *Plant Mol. Biol.* **24**: 701–713.

Fowler, D.B. and Gusta, L.V. (1979) Selection for winterhardiness in wheat. I. Identification of genotypic variability. *Crop Sci.* **19**: 769–772.

Galiba, G., Quarrie, S.A., Sutka, J., Morgounov, A. and Snape, J.W. (1995) RFLP mapping of the vernalization (*Vrn1*) and frost resistance (*Fr1*) genes on chromosome 5A of wheat. *Theor. Appl. Genet.* **90**: 1174–1179.

Gilmour, S.J., Zarka, D.G., Stockinger, E.J., Salazar, M.P., Houghton, J.M. and Thomashow, M.F. (1998) Low temperature regulation of the *Arabidopsis* CBF family of AP2 transcriptional activators as an early step in cold-induced *COR* gene expression. *Plant J.* **16**: 433–442.

Guy, C.L., Niemi, K.J. and Brambl, R. (1985) Altered gene expression during cold acclimation of spinach. *Proc. Natl Acad. Sci. USA* **82**: 3673–3677.

Hayes, P.M., Blake, T., Chen, T.H.H., Tragoonrung, S., Chen, F., Pan, A. and Liu, B. (1993) Quantitative trait loci on barley (*Hordeum vulgare* L.) chromosome 7 associated with components of winterhardiness. *Genome* **36**: 66–71.

Ingram, J. and Bartels, D. (1996) The molecular basis of dehydration tolerance in plants. *Annu. Rev. Plant Physiol. Plant Mol. Biol.* **47**: 377–403.

Jaglo-Ottosen, K.R., Gilmour, S.J., Zarka, D.G., Schabenberger, O. and Thomashow, M.F. (1998) *Arabidopsis CBF1* overexpression induces *COR* genes and enhances freezing tolerance. *Science* **280**: 104–106.

Jiang, C., Iu, B. and Singh, J. (1996) Requirement of a CCGAC *cis*-acting element for cold induction of the *BN115* gene from winter *Brassica napus*. *Plant Mol. Biol.* **30**: 679–684.

Limin, A.E., Danyluk, J., Chauvin, L.P., Fowler, D.B. and Sarhan, F. (1997) Chromosome mapping of low-temperature induced *Wcs120* family genes and regulation of cold-tolerance expression in wheat. *Mol. Gen. Genet.* **253**: 720–727.

Lin, C. and Thomashow, M.F. (1992) A cold-regulated *Arabidopsis* gene encodes a polypeptide having potent cryoprotective activity. *Biochem. Biophys. Res. Commun.* **183**: 1103–1108.

Liu, Q., Kasuga, M., Sakuma, Y., Abe, H., Miura, S., Yamaguchi-Shinozaki, K. and Shinozaki, K. (1998) Two transcription factors, DREB1 and DREB2, with an EREBP/AP2 DNA binding domain separate two cellular signal transduction pathways in drought- and low-temperature-responsive gene expression, respectively, in *Arabidopsis*. *Plant Cell* **10**: 1391–1406.

McKown, R., Kuroki, G. and Warren, G. (1996) Cold responses of *Arabidopsis* mutants impaired in freezing tolerance. *J Exp Bot* **47**: 1919–1925.

Ohme-Takagi, M. and Shinshi, H. (1995) Ethylene-inducible DNA binding proteins that interact with an ethylene-responsive element. *Plant Cell* **7**: 173–182.

Riechmann, J.L. and Meyerowitz, E.M. (1998) The AP2/EREBP family of plant transcription factors. *Biol. Chem.* **379**: 633–646.

Steponkus, P.L., Uemura, M., Joseph, R.A., Gilmour, S.J. and Thomashow, M.F. (1998) Mode of action of the *COR15a* gene on the freezing tolerance of *Arabidopsis thaliana*. *Proc. Natl Acad. Sci. USA* **95**: 14570–14575.

Stockinger, E.J., Gilmour, S.J. and Thomashow, M.F. (1997) *Arabidopsis thaliana CBF1* encodes an AP2 domain-containing transcriptional activator that binds to the C-repeat/DRE, a *cis*-acting DNA regulatory element that stimulates transcription in response to low temperature and water deficit. *Proc. Natl Acad. Sci. USA* **94:** 1035–1040.

Thomashow, M.F. (1990) Molecular genetics of cold acclimation in higher plants. *Adv. Genet.* **28:** 99–131.

Thomashow, M.F. (1998) Role of cold-responsive genes in plant freezing tolerance. *Plant Physiol* **118:** 1–7.

Xin, Z. and Browse, J. (1998) *eskimo1* mutants of *Arabidopsis* are constitutively freezing-tolerant. *Proc. Natl Acad. Sci. USA* **95:** 7799–7804.

Yamaguchi-Shinozaki, K. and Shinozaki, K. (1994) A novel *cis*-acting element in an *Arabidopsis* gene is involved in responsiveness to drought, low-temperature, or high-salt stress. *Plant Cell* **6:** 251–264.

Chapter 11

Identification and molecular characterization of *CBF2* and *CBF3*, two genes from *Arabidopsis* that encode AP2 domain-containing proteins and are induced by low temperatures

Joaquin Medina and Julio Salinas

1. Background

Many plants from temperate regions can increase their freezing tolerance in response to low non-freezing temperatures (Levitt, 1980; Sakai and Larcher, 1987). This process, called cold acclimation, is very complex and involves several biochemical and physiological changes that seem to be regulated through changes in gene expression (Thomashow, 1994). In this way, genetic analyses revealed that multiple genes are involved in cold acclimation (Thomashow, 1990), and a wide number of genes whose transcript levels accumulate in response to low temperatures have been isolated and characterized (see Capel *et al.*, 1997, 1998; Gana *et al.*, 1997; Hong *et al.*, 1997; Hughes and Dunn, 1996; Kiyosue *et al.*, 1998; Thomashow, 1994; Urao *et al.*, 1998). The precise role that these genes play in the process of cold acclimation as well as their biological function, however, remains to be determined.

   During the last years, a major goal when studying low temperature-induced gene expression has been to determine the *cis*-acting sequences that regulate such an expression. Yamaguchi-Shinozaki and Shinozaki (1994) were the first in identifying that two 9-bp DNA elements containing the core sequence CCGAC, present in the promoter of the *Arabidopsis thaliana RD29A* gene, induced low temperature gene expression when fused to a reporter gene. The low temperature-responsive element (LTRE) CCGAC, which has also been named C-repeat (CRT) by Thomashow and colleagues (Baker *et al.*, 1994), is found in one to multiple copies in the promoters of most cold-induced plant genes including the *Arabidopsis gene COR15A* (Baker *et al.*, 1994), the *Brassica napus* gene *BN115* (Jiang *et al.*, 1996) and the wheat gene *WCS120* (Ouellet *et al.*, 1998). The two DNA elements containing the CCGAC core identified by Yamaguchi-

Plant Responses to Environmental Stress, edited by M.F. Smallwood, C.M. Calvert and D.J. Bowles.

Shinozaki and Shinozaki (1994) also conferred transcription in response to dehydration stress and were designated DRE. Although many of the changes in gene expression that take place during the process of cold acclimation are mediated by abscisic acid (ABA) (Bray, 1993; Welin *et al*., 1994), the results of Yamaguchi-Shinozaki and Shinozaki (1994) revealed that the LTRE is not responsive to ABA suggesting that it imparts cold and dehydration-regulated gene expression through an ABA-independent pathway.

A further step toward the understanding of the molecular mechanisms that control the process of cold acclimation and, therefore, how low temperatures regulate gene expression was the isolation and characterization of a cDNA from *Arabidopsis* encoding a C-repeat/DRE/LTRE binding protein, CBF1 (C-repeat/DRE/LTRE binding factor, Stockinger *et al*., 1997). *CBF1* was described as a single or low copy number gene which expression levels do not change appreciably in plants exposed to low temperatures or water stress. The deduced CBF1 amino acid sequence indicated that the protein has in its N-terminal region a potential nuclear localization sequence followed by an AP2 DNA binding motif, and an acidic C-terminal half that might act as an activator domain (Stockinger *et al*., 1997). Furthermore, expression analyses in yeast demonstrated that CBF1 can function as transcriptional activator since it activates the transcription of a reporter gene containing the LTRE as an activator sequence. Recently, Jagglo-Ottosen *et al*. (1998) have shown that the overexpression of *CBF1* in *Arabidopsis* transgenic plants induces the expression of some cold-regulated genes and increases the freezing tolerance of non-cold-acclimated transgenic plants. From these results, they concluded that CBF1 can be a positive regulator of the cold acclimation process, controlling gene expression and promoting freezing tolerance.

Taking into consideration the elevated number of genes whose expression is induced during the process of cold acclimation (Capel *et al*., 1997, 1998; Gana *et al*., 1997; Hong *et al*., 1997; Hughes and Dunn, 1996; Kiyosue *et al*., 1998; Kurkela and Franck, 1990; Thomashow, 1994; Urao *et al*., 1998), it was predicted that *CBF1* homologues existed which, as in the case of the G-box and bZIP proteins (Williams *et al*., 1992), could recognize different variants of the CCGAC sequence or different LTREs depending on their surrounding sequences. Here, we describe the identification and characterization of two genes from *Arabidopsis* showing high homology with *CBF1*. These genes, we have called *CBF2* and *CBF3*, are organized in tandem with *CBF1* on chromosome 4 of *Arabidopsis* constituting a small gene family. As *CBF1*, both genes also encode proteins containing putative nuclear localization signals, AP2 DNA binding motifs and potential acidic activation domains. We show that the expression of *CBF1*, *CBF2* and *CBF3* is induced very early during the process of cold acclimation in a transient manner. However, this expression is not induced by ABA or dehydration treatments suggesting that these genes do not participate in the plant responses to ABA or water stress. Based on these results, the potential roles of *CBF2* and *CBF3* in cold acclimation and freezing tolerance are discussed.

2.   Genomic arrangement and sequence analysis of *CBF* genes

Genomic sequence analyses indicated that CBF1, CBF2 and CBF3 do not show introns among their coding sequences and are organized in tandem on chromosome 4 of *Arabidopsis*, their disposition being *CBF1*, *CBF2* and *CBF3* from 5' to 3', respectively. The high similarity existing between *CBF1*, *CBF2* and *CBF3*, together with their close linkage and the fact that they have the same transcriptional orientation clearly suggest a common origin, probably by two consecutive duplications of an ancestral gene and subsequent divergence through mutations. Interestingly, the same origin has been proposed so far for the members of five families of low temperature-responsive genes from *Arabidopsis*, in which the homologous genes are arranged in tandem in the genome. These families are *KIN1* and *COR6.6/KIN2* (Gilmour *et al.*, 1992; Kurkela and Borg-Franck, 1992), *COR15A* and *COR15B* (Wilhelm and Thomashow, 1993), *LTI78/RD29A/COR78* and *LTI65/RD29B* (Horvath *et al.*, 1993; Nordin *et al.*, 1991; Yamaguchi-Shinozaki and Shinozaki, 1993), *LTI45/LTI29* and *COR47* (Welin *et al.*, 1994), and *RCI2A* and *RCI2B* (Capel *et al.*, 1997). The reason why low temperature-regulated genes are so frequently organized in tandem remains unknown.

The 5' regulatory sequences of *CBF1*, *CBF2* and *CBF3* have diverged more than the coding regions but still keep a high level of similarity which might be the cause of the identical expression patterns shown by these genes. In this way, we identified sequence motifs that could be essential for the regulation of their expression. The sequence CAN-NTG, repeated several times in the upstream regions of the three *CBF*s, is present in the promoters of many genes that are regulated by ABA (Busk and Pagés, 1998; Guiltinan *et al.*, 1990; Williams *et al.*, 1992). The fact that *CBF1*, *CBF2* and *CBF3* are not responsive to ABA (see below) seems to indicate that this sequence does not function as an ABA-responsive element in the *CBF* promoters. The consensus LTRE core sequence, CCGAC, which has been reported to be essential for the low temperature-responsiveness of several genes (Baker *et al.*, 1994; Jiang *et al.*, 1996; Ouellet *et al.*, 1998; Yamaguchi-Shinozaki and Shinozaki, 1994) is not found in the *CBF* promoters. The question therefore remains whether the LTRE variants or the CAGCC motifs mediate the cold response shown by the *CBF* genes.

CBF1, CBF2 and CBF3 polypeptides contain a 60 amino acid motif, the AP2 domain, that is evolutionary conserved in plants (Okamuro *et al.*, 1997; Weigel, 1995) and acts as a DNA-binding domain (Buttner and Singh, 1997; Ohme-Takagi and Shinshi, 1995). In this way, it has been proposed that the binding of CBF1 to the LTRE involves its AP2 domain (Stockinger *et al.*, 1997). Furthermore, CBF2 and CBF3 also have potential nuclear localization sequences (Raikhel, 1992) and acidic C-terminal halves. In CBF1, the acidic C-terminal half has been proposed to function as transcriptional activators in yeast (Stockinger *et al.*, 1997). The high degree of similarity that exists all along the three CBF proteins (>84%), suggest that CBF2 and CBF3 probably also can bind to the LTRE and are able to activate transcription through this activator sequence. Moreover, CBF1, CBF2 and CBF3 show potential recognition sites for protein kinase C and caseine kinase II. Interestingly, some of them are conserved among the three proteins, and one of these, the Ser 59, is located into the AP2 domain. Recently, Vazquez-Tello *et al.* (1998) have proposed that the expression of *WCS120*, a low temperature-inducible gene from wheat, which contains two LTRE in its promoter region, may be regulated by nuclear factors whose binding activity is modulated by phosphorylation/dephosphorylation mechanisms. We speculate that the potential phos-phorylation sites found in the CBF proteins may play important roles in their ability to bind LTREs and, therefore, to modulate gene expression.

3. Regulation of *CBF* gene expression

Expression analyses revealed that *CBF2* and *CBF3* are positively regulated by low tem-perature. Furthermore, contrary to what was described first by Stockinger *et al.* (1997), in our experimental conditions *CBF1* transcripts also accumulate in response to low temperature. The cold-inducible expression of *CBF* genes do not show marked differ-ences. It is not organ-specific, since CBF mRNAs accumulate to similar levels in differ-ent organs of *Arabidopsis*, and, in contrast to most cold-regulated genes characterized in *Arabidopsis* (Hughes and Dunn, 1996; Thomashow, 1994) is transient. The accumula-tion of *CBF* transcripts is very quick after transferring plants to low temperature condi-tions, reaching maximal levels after 1 h of exposure and decreasing thereafter. This expression pattern suggests that *CBF* genes should be involved in responses that are transiently produced when plants are exposed to low temperatures, and fits well with the notion that their induction should be an early amplification event in the low tempera-ture-induced signalling cascade, preceding and prompting the accumulation of tran-scripts corresponding to *CBF*-regulated genes.

The expression of CBFs is not regulated by ABA or water stress, two treatments that have been shown to increase plant freezing tolerance (Chen and Gusta, 1983; Cloutier and Simikovitch, 1982; Mäntylä *et al.*, 1995) and induce the expression of most cold-inducible genes (Capel *et al.*, 1997; Hong *et al.*, 1997; Hughes and Dunn, 1996; Kirch *et al.*, 1997; Rouse *et al.*, 1996; Shinozaki and Yamaguchi-Shinozaki, 1996; Thomashow, 1994). That CBF genes are not responsive to ABA is consistent with the fact that they are involved in regulating the expression of low temperature-inducible genes through an ABA-independent pathway. However, since LTRE seems to be respon-sive to dehydration (Yamaguchi-Shinozaki and Shinozaki, 1994), the fact that the expression of CBFs is not induced by water stress suggests that other LTRE-binding proteins than CBFs may mediate dehydration-regulated gene expression through LTREs in an ABA-independent pathway.

4. Future directions

CBF1 has been described as a transcriptional activator that binds to the LTRE sequence inducing the expression of some low temperature genes and increasing freezing toler-ance (Jagglo-Ottosen *et al.*, 1998; Stockinger *et al.*, 1997). Our data indicate that *CBF2* and *CBF3* have an astonishing degree of homology with *CBF1*, and show an identical expression pattern. This suggests that they may fulfil a similar function to *CBF1* con-trolling the level of low temperature-regulated gene expression, and promoting freezing tolerance. We hypothesize that, given the large number of genes whose expression is induced in response to low temperature, differences in the sequences of the CCGAC core element, or/and in the sequences that surround it, might result, as described for the G-box sequence CANNTG and the bZIP proteins (Williams *et al*., 1992), in the recruitment of distinct CBF proteins. Having available the *CBF* genes makes it possible to carry out a number of *in vivo* and *in vitro* experiments that will allow this hypothesis to be tested.

References

Baker, S.S., Wilhelm, K.S. and Thomashow, M.F. (1994) The 5′-region of *Arabidopsis thaliana cor15a* has *cis*-acting elements that confer cold-, drought- and ABA-regulated gene expression. *Plant Mol. Biol.* **24:** 701–713.

Bray, E.A. (1993) Molecular responses to water deficit. *Plant Physiol* **103:** 1035–1040.

Busk, K.P. and Pages, M. (1998) Regulation of abscisic acid-induced transcription. *Plant Mol. Biol.* **37:** 425–435.

Buttner, M. and Singh, K.B. (1997) *Arabidopsis thaliana* ethylene-responsive element binding protein (AtEBP), an ethylene-inducible, GCC box DNA-binding protein interacts with an ocs element binding protein. *Proc. Natl Acad. Sci. USA* **94:** 5961–5966.

Capel, J., Jarillo, J.A., Salinas, J. and Martinez-Zapater, J.M. (1997) Two homologous low-temperature-inducible genes from *Arabidopsis* encode highly hydrophobic proteins. Plant Physiol **2:** 569–576.

Capel, J., Jarillo, J.A., Madueño, F., Jorquera, M.J., Martinez-Zapater, J.M. and Salinas, J. (1998) Low temperature regulates *Arabidopsis* Lhcb gene expression in a light-independent manner. *Plant J.* **13:** 411–418.

Chen, T.H.H. and Gusta, L.V. (1983) Abscisic acid induced freezing resistance in cultured plant cells. *Plant Physiol.* **73:** 71–75.

Cloutier, Y. and Simikovitch, D. (1982) Correlation between cold- and drought-induced frost hardiness in winter wheat and rye varieties. *Plant Physiol.* **69:** 256–258.

Gana, J.A., Sutton, F. and Kenefick, D.G. (1997) cDNA structure and expression patterns of a low-temperature-specific wheat gene *tacr7*. *Plant Mol. Biol.* **4:** 643–650.

Gilmour, S.J., Artus, N.N. and Thomashow, M.F. (1992) cDNA sequence analysis and expression of two cold regulated genes of *Arabidopsis thaliana*. *Plant Mol. Biol.* **18:** 13–21.

Guiltinan, M.J., Marcotte, W.R. and Quatrano, R.S. (1990) A plant leucine zipper protein that recognizes an abscisic responsive element. *Science* **250:** 267–271.

Hong, S.W., Jon, J.H., Kwak, J.M. and Nam, H.G. (1997) Identification of a receptor-like protein kinase gene rapidly induced by abscisic acid, dehydration, high salt, and cold treatments in *Arabidopsis thaliana*. *Plant Physiol.* **4:** 1203–1212.

Horvath, D.P., McLarney, B.K. and Thomashow, M.F. (1993) Regulation of *Arabidopsis thaliana* L. (Heyn) *cor78* in response to low temperature. *Plant Physiol.* **103:** 1047–1053.

Hughes, M.A. and Dunn, M.A. (1996) The molecular biology of plant acclimation to low temperature. *J. Exp. Bot.* **47:** 291–305.

Jaglo-Ottosen, K.R., Gilmour, S.J., Zarka, D.G., Schabenberger, O. and Thomashow, M.F. (1998) *Arabidopsis CBF1* overexpression induces *COR* genes and enhances freezing tolerance. *Science* **280:** 104–106.

Jiang, C., Iu, B. and Singh, J. (1996) Requirement of a CCGAC cis-acting element for cold induction of the BN115 gene from winter *Brassica napus*. *Plant Mol. Biol.* **30:** 679–684.

Kirch, H.H, van Berkel, J., Glaczinski, H., Salamini, F. and Gebhardt, C. (1997) Structural organization, expression and promoter activity of a cold-stress-inducible gene of potato. *Plant Mol. Biol.* **5:** 897–909.

Kiyosue, T., Abe, H., Yamaguchi-Shinozaki, K. and Shinozaki, K. (1998) *ERD6*, a cDNA clone for an early dehydration-induced gene of *Arabidopsis*, encodes a putative sugar transporter. *Biochim. Biophys. Acta* **1370:** 187–191.

Kurkela, S. and Franck, M. (1990) Cloning and characterization of a cold- and ABA induced Arabidopsis gene. *Plant Mol. Biol.* **15:** 137–144.

Kurkela, S. and Borg-Franck, M. (1992) Structure and expression of *Kin2*, one of two cold- and ABA-induced genes of *Arabidopsis thaliana. Plant Mol. Biol.* **19:** 689–692.

Levitt, J. (1980) *Responses of Plants to Environmental Stresses:* Chilling, Freezing and High Temperature Stresses. Academic Press, New York.

Mäntylä, E., Lang, V. and Palva, T. (1995) Role of abscisic acid in drought-induced freezing tolerance cold acclimation, and accumulation of LTI78 and RAB18 proteins in *Arabidopsis thaliana. Plant Physiol.* **107:** 141–148.

Nordin, K., Heino, P. and Palva, E.T. (1991) Separate signal pathways regulate the expression of a low-temperature-induced gene in *Arabidopsis thaliana* (L.) Heynh. *Plant Mol. Biol.* **187:** 169–183.

Ohme-Tagaki, M. and Shinshi, H. (1995) Ethylene-inducible DNA binding proteins that interact with an ethylene-responsive element. *Plant Cell* **7:** 173–182.

Okamuro, J.K., Caster, B., Villarroel, R., Van Montagu, M. and Jofoku, D.K. (1997) The AP2 domain of APETALA2 defines a large new family of DNA binding proteins in Arabidopsis. *Proc. Natl Acad. Sci. USA* **13:** 7076–7081.

Ouellet, F., Vazquez-Tello, A. and Sarhan, F. (1998) The wheat *wcs120* promoter is cold-inducible in both monocotyledonous and dicotyledonous species. *FEBS Lett.* **423:** 324–328.

Raikhel, N. (1992) Nuclear targeting in plants. *Plant Physiol.* **100:** 1627–1632.

Rouse, D.T., Marotta, R. and Parish, R.W. (1996) Promoter and expression studies on an *Arabidopsis thaliana* dehydrin gene. *FEBS Lett.* **3:** 252–256.

Sakai, A. and Larcher, W. (1987) Frost survival of plants. In: *Responses and Adaptation to Freezing Stress.* Springer-Verlag, New York.

Shinozaki, K. and Yamaguchi-Shinozaki, K. (1996) Molecular responses to drought and cold stress. *Curr. Opinion Biotechnol.* **7:** 161–167.

Stockinger, E.J., Gilmour, S.J. and Thomashow, M.F. (1997) *Arabidopsis thaliana CBF1* encodes an AP2 domain-containing transcriptional activator that binds to the C-repeat/DRE, a cis-acting DNA regulatory element that stimulates transcription in response to low temperature and water deficit. *Proc. Natl Acad. Sci. USA* **94:** 1035–1040.

Thomashow, M.F. (1990) Molecular genetics of cold acclimation in higher plants. *Adv. Gene* **28:** 99–131.

Thomashow, M.F. (1994) *Arabidopsis thaliana* as a model for studying mechanisms of plant cold tolerance. In: *Arabidopsis* (eds E.M. Meyerowitz and C.R. Somerville). Cold Spring Harbor Laboratory Press, Cold Spring Harbor, NY, pp. 807–834.

Urao, T., Yakubov, B., Yamaguchi-Shinozaki, K. and Shinozaki, K. (1998) Stress-responsive expression of genes for two-component response regulator-like proteins in *Arabidopsis thaliana. FEBS Lett* **427:** 175–178.

Vazquez-Tello, A., Ouellet, F. and Sarhan, F. (1998) Low temperature-stimulated phosphorylation regulates the binding of nuclear factors to the promoter of *Wcs120*, a cold-specific gene in wheat. *Mol. Gen. Genet.* **257:** 157–166.

Weigel, D. (1995) The APETALA2 domain is related to a novel type of DNA binding domain. *Plant Cell* **7:** 388–389.

Welin, B.V., Olson, A., Nylander, M. and Palva, E.T. (1994) Characterization and differential expression of *dhn/lea/rab*-like genes during cold acclimation and drought stress in *Arabidopsis thaliana. Plant Mol. Biol.* **26:** 131–144.

Wilhelm, K. and Thomashow, M.F. (1993) *Arabidopsis thaliana cor l5b*, an apparent homologue of *cor l5a*, is strongly responsive to cold and ABA, but not drought. *Plant Mol. Biol.* **23:** 1073–1077.

Williams, M.E., Foster, R. and Chua, N.H. (1992) Sequences flanking the hexameric G-box core CACGTG affect the specificity of protein binding. *Plant Cell* **4:** 485–496.

Yamaguchi-Shinozaki, K. and Shinozaki, K. (1993) Characterization of the expression of a desiccation-responsive *rd29* gene of *Arabidopsis thaliana* and analysis of its promoter in transgenic plants. *Mol. Gen. Genet.* **236:** 331–340.

Yamaguchi-Shinozaki. K. and Shinozaki, K. (1994) A novel cis-acting element in an Arabidopsis gene is involved in responsiveness to drought, low-temperature or high-salt stress. *Plant Cell* **6:** 251–264.

Chapter 12

Low temperature regulation of gene expression in cereals

Monica A. Hughes, Anthony P.C. Brown, Senay Vural and M. Alison Dunn

1. Background

Over-wintering cereal crops have evolved two mechanisms, namely vernalization response and low temperature acclimation, which enable them to survive low temperature stress. Both mechanisms are induced by a similar range of low positive temperatures (typically below 10°C) and genetic evidence indicates that they are controlled by interrelated genetic systems (Fowler *et al.*, 1996a; Fowler *et al.*, 1996b; Karsai *et al.*, 1997). Cytogenetic analysis shows that in wheat, for example, at least 10 of the 21 pairs of chromosomes are involved in the control of the low temperature survival traits, frost tolerance and winter hardiness (Galiba *et al.*, 1998). Low temperature or frost acclimation has been shown to be accompanied by altered gene expression, and a number of genes which are upregulated at the mRNA level have been isolated (Hughes and Dunn, 1996). These low temperature-responsive (LTR) genes are induced by the low positive temperatures that both acclimate and vernalize winter cereals; they were primarily identified by their differential expression (Hughes and Dunn, 1996).

The two primary questions that apply to these LTR genes are:

(i)   How does temperature control LTR gene expression?
(ii) What is the function of the LTR gene product(s) and does this have a role in either low temperature acclimation or vernalization?

This chapter is concerned with the control of gene expression and there are a number of key questions that can be asked about low temperature gene regulation:

(i)   How do plants sense the low temperature and what is the nature of the signal transduction pathway(s) that control gene expression? A number of putative components in low temperature signal transduction have been identified, and these include abscisic acid (Dörffling *et al.*, 1998), calcium (Knight *et al.*, 1996) and phosphorylation (Vazquez-Tello *et al.*, 1998). What is the role of these potential components of a signal transduction cascade and how do they interact?
(ii) Since most of the genes that have been isolated are identified by increased steady state mRNA levels at low temperature, are these genes controlled at a transcriptional or post-transcriptional level?

Plant Responses to Environmental Stress, edited by M.F. Smallwood, C.M. Calvert and D.J. Bowles.
© 1999 BIOS Scientific Publishers, Oxford.

89

(iii) Does post-transcriptional control involve mRNA stabilizing factors at low temperature or destabilizing factors at normal (ambient) temperatures?
(iv) What is the nature of the stabilizing/destabilizing factors and how do they interact with LTR mRNA?
(v) Does transcriptional control involve the same promoter elements in all genes?
(vi) What is the nature of the transcriptional complex(es)?
(vii) Does chromatin structure play a part in the control of LTR gene expression, and if so, what factors are involved in chromatin remodelling at low temperature?

Most of the key questions that apply to the low temperature regulation of gene expression in cereals also apply to other plants. However, the bottlenecks that hinder progress are largely specific to cereals. The major bottleneck must be the lack of an efficient cereal transformation technique that can be used for the experimental analysis of cereal genes. This problem is followed by the large size and complexity of the genomes of those cereals that vernalize and frost acclimate. The cereal which has a small genome and is considered as a model species for the group by many workers, is rice; but rice is not a good model for studying low temperature responses. In addition, although considerable variation in low temperature response exists within individual cereal species, in general this has not been well characterized either physiologically or genetically. Furthermore, because the material is distributed very broadly within the international plant breeding community, it can be difficult to acquire.

In contrast to other plant species, the inbreeding genetic system of many cereals means that it is possible to produce large quantities of uniform material from these plants. This together with a long tradition of their use in biochemical studies, makes them suited to a number of biomolecular or biochemical studies which are difficult in other plants. These experimental approaches have been used (in combination with other studies) to analyse the control of LTR genes in wheat and barley and, although these studies are at a relatively early stage, they do indicate that when developed in conjunction with the powerful suite of available yeast-based cloning techniques, considerable progress can be expected in the near future.

2. Low temperature control of gene expression

The pattern of expression of a large number of LTR cereal genes has been studied using northern and/or western blotting (see, for example, Crosatti et al., 1996; Grossi et al., 1998). However, there are only a relatively small number of studies of the molecular mechanism(s) of low temperature control of gene expression (*Table 1*).

In a study of winter barley genes that were isolated by differential screening of a shoot meristem cDNA library, nuclear run-on transcription analysis was used to determine whether a transcriptional or a post-transcriptional mechanism was the primary determinant of elevated steady-state mRNA levels at low temperature (Dunn et al., 1994). The results of this analysis show that four genes (*blt4.9, blt101, blt1015* and *blt410*) have high levels of transcript produced by low temperature nuclei but very low levels by the control temperature nuclei, indicating that these genes are transcriptionally

Table 1. Winter cereal low temperature-responsive (LTR) genes

Gene family	Species	Location (chromosome number)	Control	Reference
blt4	Winter barley (cv Igri)	3	Transcriptional	Dunn *et al.* (1998)
blt14	Winter barley (cv Igri)	2	Post-transcriptional	Phillips *et al.* (1997)
blt101	Winter barley (cv Igri)	4	Transcriptional	Dunn *et al.* (1994)
Wcs120	Winter wheat (cv Fredrick)	6 A, B, D	Transcriptional	Vazquez-Tello *et al.* (1998)

regulated in response to low temperature. Two further genes (*blt 63* and *blt 49*) have high levels of their respective transcripts at the control temperature, both of which increase markedly with the low temperature treatment, indicating that these genes are also transcriptionally regulated at low temperature.

In contrast to this low temperature transcriptional response, two genes (*blt14* and *blt411*) have very low levels of detectable transcript in both temperature treatments, indicating low transcription rates at both low and control temperatures. One gene (*blt801*) which also accumulates high steady-state levels at low temperature, appears to be transcriptionally down-regulated in the cold, although transcription rates of this gene are higher at both temperatures compared to those of *blt14* or *blt411*. Since all three transcripts (*blt14*, *blt411* and *blt801*) accumulate at low temperature but not the control temperature, these results suggest that their regulation by temperature is post-transcriptional. This study indicates that one third of the barley genes investigated are primarily regulated by a post-transcriptional control mechanism.

2.1 Post-transcriptional control

The molecular mechanisms of post-transcriptional control of gene expression are not well understood but they have been shown to be important in a number of studies of the response of plants to environmental change (Gallie, 1993). Stability of mRNA from the barley gene *blt14.0* during cold acclimation and de-acclimation was investigated by gel blot analysis using metabolic inhibitors (Phillips *et al.*, 1997). Since global effects of such inhibitors can give misleading results, the same northern blots were also analysed for mRNA levels of the transcriptionally controlled barley LTR gene, *blt101.1* (Goddard *et al.*, 1993).

Cordycepin, which inhibits plant RNA synthesis when used at high concentrations, has been successfully used in plant systems to analyse the stability of transcripts. The relationship between *blt14.0* mRNA stability and translation was also investigated using cycloheximide. In eukaryotes, cycloheximide stabilizes most unstable transcripts, either through arresting translation of the transcript itself (a *cis* effect) or by preventing translation of an unstable *trans*-acting factor required for mRNA degradation.

The mRNA stability of pre-existing *blt14.0* mRNA was studied in barley shoot explants, in the presence or absence of 800 μM cordycepin or 70 μM cycloheximide in de-acclimation and acclimation environments. The same northern blots were analysed for mRNA levels of *blt101.1*. Differences between the effect of these inhibitors on *blt14.0* and *blt101.1* mRNA levels are shown in *Figure 1*. Levels of mRNA were measured, over a 60 h period. *Figure 1* shows that maintenance of high *blt14.0* steady-state mRNA levels at low temperature is impaired by both cordycepin and cycloheximide. In contrast, although cordycepin reduces *blt101.1* mRNA steady-state levels at this temperature, cycloheximide treatment causes a slight overall increase (Phillips *et al.*, 1997).

The markedly reduced levels of *blt14.0* mRNA in the presence of cordycepin and cycloheximide in a low temperature environment, indicates that both transcription and translation are required to maintain *blt14.0* steady-state mRNA levels at 2°C. These data suggest that a stabilizing protein factor(s), which is critical for the increase in *blt14.0*

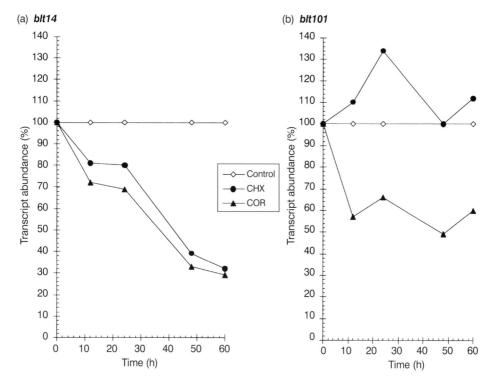

Figure 1. Analysis of etiolated barley (cv Igri) shoot (maintained at 2°C) steady-state mRNA levels in the presence of 800 μm cordycepin or 70 μm cycloheximide. Seedlings were grown in the dark at 22°C for 4 days followed by 2°C for a further 7 days. Seedling shoots were excised and incubated at 22°C for 2 h in the absence or presence of cordycepin (COR) or cycloheximide (CHX), then returned to 2°C. Shoot bases were harvested at 0, 12, 24, 48 and 60 h. (a) *blt14.0*. (b) *blt101*. The densitometric data has been adjusted for loading using a 5.8S rDNA probe and is normalized relative to the value obtained from control seedlings (Control). Reprinted from *Plant Molecular Biology*, vol. 33, 1997, pp. 1013–1023, mRNA stability and localisation of the low temperature-responsive barley gene family *blt14*, Phillips, J. R., Dunn, M.A. and Hughes, M.A., fig 7, with kind permission from Kluwer Academic Publishers.

steady-state mRNA levels, is transcribed and translated at low temperature. *blt14.0* steady-state mRNA levels are not significantly affected by the presence of cordycepin at 18°C, whereas in the presence of cycloheximide a dramatic reduction of *blt14.0* steady-state mRNA levels is observed. This also suggests that the decay of *blt14.0* mRNA during de-acclimation is controlled by a labile protein factor(s) which is not synthesized in the presence of cycloheximide.

2.2 *Transcriptional control: promoter element interaction with nuclear proteins*

The *cis*-acting promoter elements that control elevated transcription rates at low positive temperatures have been investigated in members of three cereal gene families (*blt4* and *blt101* in barley and *Wcs120* in wheat). The *blt4* barley gene family encodes non-specific lipid transfer proteins and has been shown, by *in situ* localization, to be expressed in the epidermal cells of leaves. The transcriptionally controlled LTR member of this gene family, *blt4.9*, is predominantly expressed in shoot meristems. The promoter region (1938 bases) of *blt4.9* contains sequence motifs which have been implicated in low temperature, abscisic acid and other environmental factors (White *et al.*, 1994). Deletion analysis showed that a 42-base sequence proximal to, but not including, the CAAT and TATA boxes, conferred enhanced low temperature response to a reporter gene in a barley shoot explant transient expression system. Electrophoretic mobility shift analysis (EMSA) was used with nuclear proteins from either low temperature or control temperature treated plants to further investigate the *blt4.9* promoter. This identified a hexanucleotide, CCGAAA, within the 42-base LTR promoter region, as the

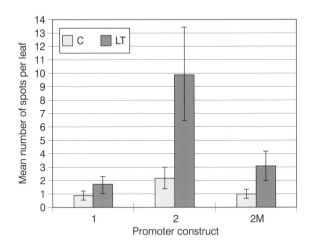

Figure 2. Transient reporter gene expression analysis of mutant LTRE-1. Transient β-glucuronidase gene expression in barley leaf explants following particle bombardment with *blt4.9* promoter constructs 1 (−175 bases), 2 (−216 bases) and 2M (−216 bases with LTRE-1, CCGAAA, changed to ATATAA). Expression is shown as the mean number of blue spots per leaf. Error bars, standard error; C, control explants; LT, low temperature-treated explants. Reprinted from *Plant Molecular Biology*, vol. 38, 1998, pp. 551–564, Identification of promoter elements in a low temperature-responsive gene (blt4.9) from barley, Dunn, M.A., White, A.J., Vural, S. and Hughes, M.A., fig 8, with kind permission from Kluwer Academic Publishers.

binding site of a low mobility nuclear protein complex. This complex was present in nuclear extracts from both low temperature-treated and control plants and was the only complex formed within this region. Mutation of the CCGAAA motif within the LTR 42-base promoter sequence reduced low temperature responsiveness to basal levels (*Figure 2*) (Dunn *et al.*, 1998).

Several studies of dicotyledon species (*Arabidopsis*: Baker *et al.*, 1994; Yamaguchi-Shinozaki and Shinozaki, 1994 and *Brassica napus*: Jiang *et al.*, 1996) have identified a conserved pentanucleotide (CCGAC) as a LTR element, and a transcription factor (CBF1) which binds to this element has been identified in *Arabidopsis* (Stockinger *et al.*, 1997). This dicotyledonary element is clearly related to the *blt4.9* barley gene putative LTR element, CCGAAA. However, recent work shows that the *Arabidopsis* CBF1 transcription factor (Stockinger *et al.*, 1997) expressed in *E. coli* using the pET28a vector (Novagen Inc.), does not bind the barley CCGAAA putative LTR element (S. Kural, M.A. Dunn and M.A. Hughes, unpublished data). A CCGAC element also exists in the barley *blt4.9* promoter upstream of the CCGAAA element. However, EMSA indicates that this element is not a binding site for nuclear factors.

The promoter of the barley gene, *blt101.1*, has also been investigated by a combination of transient expression of promoter deletion and mutated reporter gene constructs, delivered to barley leaves on gold particles via Biolistics, as well as EMSA of nuclear protein binding to synthetic oligonucleotides. Unlike *blt4.9*, *blt101.1* is only responsive to low temperature and is not induced by abscisic acid or drought. The *blt101.1* promoter does not contain either a CCGAAA or a CCGAC element. An element, which binds nuclear proteins from both low temperature and normal temperature-treated plants, has been identified within a 42-base fragment of the 104 LTR base proximal region of the promoter. However, mutations of this element do not lead to the loss of the LT response in transient expression assays (A.P.C. Brown, M.A. Dunn and M.A. Hughes, unpublished data). It is expected that an LTR element resides elsewhere in the 104-base proximal region of the *blt101.1* promoter.

The *Wcs120* gene from wheat is a member of a family of related genes and is specifically regulated by low temperature. The accumulation of *Wcs120* mRNA and protein has been shown to correlate closely with the differential capacity of wheat cultivars to develop freezing tolerance (Limin *et al.*, 1995). The *Wcs120* promoter contains a number of potential control elements, including the CCGAC pentanucleotide at -175 and -337 bases (Vazquez-Tello *et al.*, 1998). EMSA analysis of 860 bases of the *Wcs120* promoter subdivided into six overlapping fragments between 100 and 160 bases long, shows a complex pattern of nuclear protein interaction (Vazquez-Tello *et al.*, 1998), which compares with a similar study of the barley *blt4.9* promoter (Dunn *et al.*, 1998). Interestingly, Vazquez-Tello *et al.* (1998) (unlike the barley gene study) showed that nuclear proteins from low temperature-acclimated plants were unable to bind to these promoter fragments unless the nuclear protein extract had been dephosphorylated with alkaline phosphatase. This result was paralleled by increased levels of both Ca^{2+}-dependent and Ca^{2+}-independent kinase activities in low temperature-acclimated nuclear extracts and the *in vivo* stimulation of the *Wcs120* protein family in plants treated with the phosphatase inhibitor, okadaic acid (Vazquez-Tello *et al.*, 1998). This result is unexpected since it implies that low temperature stimulation of *Wcs120* expression is due to inactivation of DNA-binding factors which interact with all six fragments of the promoter. No putative positive acting factor was seen in this study.

The identification of LTR elements within the promoters of transcriptionally controlled LTR cereal genes will allow the isolation (via yeast one-hybrid cloning systems) of genes encoding transcription factors which bind these elements. This technique has already been successfully used to isolate the *Arabidopsis* transcription factor gene, *CBF1* (Stockinger *et al.*, 1997).

2.3 Transcriptional control: chromatin structure

In order to determine whether local changes in promoter chromatin structure may contribute to changes in gene expression at low temperature, a PCR-based method was used to analyse the chromatin structure of the active and inactive promoter of *blt101.1*. As a comparative control, the constitutive promoter of the post-transcriptionally controlled gene *blt14.1* was also analysed (Phillips *et al.*, 1997). Intact nuclei were isolated from the shoot meristematic tissue of control and low temperature-treated tissue, and the released chromatin subjected to degradation by either micrococcal nuclease or DNaseI. The integrity of defined promoter sequences in digested samples was then examined by PCR amplification.

The study of *blt101.1* in these experiments encompassed two overlapping regions of the promoter, both of which contain the 42 base pair sequence used in EMSAs. Promoter sequences from *blt14.1* were from an equivalent region of the *blt14.1* promoter (-230 to -1) and gave approximately the same sized PCR fragment.

The difference in sensitivity of intact chromatin to nucleases, reflected in the differ-

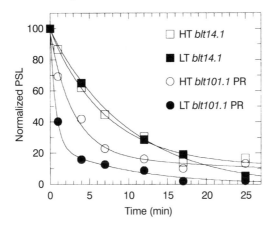

Figure 3. Nuclease (Dnase1) sensitivity of *blt101.1* and *blt14.1* promoters in intact chromatin isolated from control (HT) and low temperature (LT) treated barley leaves. Graph showing degradation rates of the *blt14.1* promoter region -230 to -1 bases and the *blt101.1* promoter proximal region (PR) -261 to -2 bases. For each primer pair, PCR conditions were adjusted such that the template concentration was limiting, in order to produce a quantitative reaction. PCR products were electrophoresed, blotted, probed with the appropriate promoter sequence and quantified using a FujiFilm BAS-1500 bioimage analyser.

ing rates of decay of the two genes, is illustrated in *Figure 3*. The average ratio (four experiments) of the initial rates of decay between control and low temperature-treated tissue for the two regions of *blt10.1* are 1:1.56 and 1:1.9, respectively, indicating that these regions become more susceptible to nuclease digestion upon cold induction. In contrast, the ratio for the post-transcriptionally controlled *blt14.1* gene is 1:0.96, indicating little change in sensitivity to nucleases at low temperature, as expected for a promoter whose state of transcriptional activity remains unaltered by temperature.

It is clear from the existing cereal studies in wheat and barley, that a single molecular mechanism of transcriptional control for cereal LTR genes is unlikely. Furthermore, evidence from the barley gene, *blt101.1*, indicates that chromatin remodelling may also be involved in the low temperature control of gene expression.

References

Baker, S.S., Wilhelm, K.S. and Thomashow, M.F. (1994) The 5′-region of *Arabidopsis thaliana cor15a* has *cis*-acting elements that confer cold-regulated, drought-regulated and ABA-regulated gene expression. *Plant Mol. Biol.* **24**: 701–713.

Crosatti, C., Nevo, E., Stanca, A.M. and Cattivelli, L. (1996) Genetic analysis of the accumulation of COR14 proteins in wild (*Hordeum spontaneum*) and cultivated (*Hordeum vulgare*) barley. *Theor. Appl. Genet.* **93**: 975–981.

Dörffling, K., Abromeit, M., Bradersen, U., Dörffling H. and Melz, G. (1998) Involvement of abscisic acid and proline in cold acclimation of winter wheat. In: *Plant Cold Hardiness* (eds P.H. Li and T.H.H. Chen). Plenum Press, New York, pp. 283–292.

Dunn, M.A., Goddard N.J., Zhang L., Pearce R.S. and Hughes M.A. (1994) Low temperature-responsive barley genes have different control mechanisms. *Plant Mol. Biol.* **24**: 879–888.

Dunn, M.A., White, A.J., Vural, S. and Hughes, M.A. (1998) Identification of promoter elements in a low temperature-responsive gene (blt4.9) from barley. *Plant Mol. Biol.* **38**: 551–564

Fowler, D.B., Chauvin, L.P., Limin, A.E. and Sarhan, F. (1996b) The regulatory role of vernalisation in the expression of low temperature-induced genes in wheat and rye. *Theor. Appl. Genet.* **93**: 554–559.

Fowler, D.B., Limin, A.E, Wang, S.Y. and Ward, R.W. (1996a) Relationship between low temperature tolerance and vernalisation response in wheat and rye. *Can. J. Plant Sci.* **76**: 37–42.

Galiba, G., Kerepesi, I., Snape, J.W. and Sutka, J. (1998) Mapping of genes controlling cold hardiness on wheat 5A and its homologous chromosomes of cereals. In: *Plant Cold Hardiness* (eds P.H. Li, and T.H.H. Chen). Plenum Press, New York, pp. 89–98.

Gallie, D.R. (1993) Post-transcriptional regulation of gene expression in plants. *Annu. Rev. Plant Physiol. Plant Mol. Biol.* **44**: 77–105.

Grossi, M., Giorni, E., Rizza, F., Stanca, A.M. and Cattivelli, L. (1998) Wild and cultivated barleys show differences in the expression pattern of a cold-regulated gene family under different light and temperature conditions. *Plant Mol. Biol.* **38**: 1061–1069.

Hughes, M.A. and Dunn, M.A. (1996) The molecular biology of plant acclimation to low temperature. *J. Exp. Bot.* **47**: 291–305.

Jiang, C., Iu, B. and Singh, J. (1996) Requirement of a CCGAC *cis*-acting element for cold induction of *BN*115 gene from *B. napus*. *Plant Mol. Biol.* **30**: 679–684.

Karsai, A.U., Meszaros, K., Bedo, Z., Hayes, P.M., Pan, A. and Chen, F. (1997) Genetic analysis of the components of winter hardiness in barley (*Hordeum vulgare* L.). *Acta Biol. Hungarica* **48**: 67–76.

Knight, H., Trewavas, A.J. and Knight, M.R. (1996) Cold calcium signalling in *Arabidopsis* involves two cellular pools and a change in calcium signature after acclimation. *Plant Cell* **8**: 489–503.

Limin, A.E., Houde, M, Chauvin, L.P., Fowler, D.B. and Sarhan F. (1995) Expression of the cold-induced wheat gene *Wcs120* and its homologues in related species and interspecific combinations. *Genome* **38:** 1023–1031.

Phillips, J.R., Dunn, M.A. and Hughes, M.A. (1997) mRNA stability and localisation of the low temperature-responsive barley gene family *blt14*. *Plant Mol. Biol.* **33:** 1013–1023.

Stockinger, E.J., Gilmour, S.J. and Thomashow, M.F. (1997) *Arabidopsis thaliana CBF1* encodes an AP2 domain-containing transcriptional activator that binds to the C-repeat/DRE, a *cis*-acting DNA regulatory element that stimulates transcription in response to low temperature and water deficit. *Proc. Natl Acad. Sci. USA* **94:** 1035–1040

Vazquez-Tello, A., Ouellet, F. and Sarham, F. (1998) Low temperature-stimulated phosphorylation regulates the binding of nuclear factors to the promoter of *Wcs120*, a cold-specific gene in wheat. *Mol. Gen. Genet.* **257:** 157–166.

White, A.J., Dunn, M.A., Brown, K. and Hughes, M.A. (1994) Comparative analysis of genomic sequence and expression of a lipid transfer protein gene family in winter barley. *J. Exp. Bot.* **45:** 1885–1892.

Yamaguchi-Shinozaki, K. and Shinozaki, K. (1994) A novel *cis*-acting element in an *Arabidopsis* gene is involved in responsiveness to drought, low temperature, or high-salt stress. *Plant Cell* **6:** 251–264.

Chapter 13

The quest to elucidate the role of the *COR* genes and polypeptides in the cold acclimation process

Peter L. Steponkus and Matsuo Uemura

1. Background

Since 1985, when Guy *et al.* first reported that gene expression is altered during cold acclimation, molecular biologists have made remarkable progress in identifying an ever-increasing number of genes that are regulated by low temperatures. Some of these genes encode polypeptides of known metabolic function, and the level of expression is only increased during exposure to low temperature – though in some instances the increase is substantial. For example, in *Arabidopsis thaliana* the transcript levels for phenylammonia lyase and chalcone synthetase are significantly increased by exposure to low temperatures (Leyva *et al.*, 1995). More intriguing are the cold-regulated genes that are expressed only during exposure to low temperatures and encode polypeptides that have little or no amino acid sequence homology with known proteins. These include the *COR* genes and *LTI* genes of *A. thaliana* (Gilmour *et al.*, 1992; Lin and Thomashow, 1992a; Nordin *et al.*, 1993), the *COR* and *pao86* genes of barley (Cattivelli and Bartels, 1990; Crosatti *et al.*, 1996) and the *A/ES* genes of alfalfa (Luo *et al.*, 1992). It has been long speculated that these genes might have roles in the cold acclimation process because their level of expression is positively correlated with the freezing tolerance of several alfalfa cultivars (Mohapatra *et al.*, 1989), and the synthesis of the COR polypeptides coincides closely with increases in freezing tolerance (Guy and Haskell, 1987; Mohapatra *et al.*, 1987; Thomashow *et al.*, 1990). *COR6.6*, *COR15a*, *COR47* and *COR78* are among the genes that are most highly induced during cold acclimation of *A. thaliana* (Thomashow, 1990, 1993).

However, for more than a decade, only very indirect evidence was available to support speculation that the *COR* genes and polypeptides have a functional role in the cold acclimation process. Most of the speculation was based on very general similarities that the COR polypeptides have in common with LEA (late embryogenesis abundant) proteins and the 'cryoprotective proteins' first described by Volger and Heber (1975) and more recently by Hincha *et al.* (1996). For example, Thomashow (1993) determined that the polypeptides encoded by *COR6.6*, *COR15a*, *COR47* and *COR78* are hydrophilic and boiling stable, which is a characteristic of both LEA proteins and the

Plant Responses to Environmental Stress, edited by M.F. Smallwood, C.M. Calvert and D.J. Bowles.
© 1999 BIOS Scientific Publishers, Oxford.

'cryoprotective proteins'. Therefore, he speculated that the COR polypeptides might be homologues of the 'cryoprotective proteins'. Because the 'cryoprotective proteins' reportedly confer cryoprotection to isolated thylakoids, it is often inferred that the COR polypeptides might also have a cryoprotective function.

For several years, the only direct evidence that one of the COR polypeptides might have a cryoprotective effect was the report by Lin and Thomashow (1992b) that the COR15a polypeptide isolated from *A. thaliana* exhibits 'profound cryoprotective activity' for lactate dehydrogenase. However, subsequent studies (S.J. Gilmour and M.F. Thomashow, unpublished data) revealed that the cryoprotective effect of COR15am (the mature form of the polypeptide) is no greater than bovine serum albumin. In 1994, Uemura *et al.* reported that, unlike the 'cryoprotective proteins' described by Heber and his colleagues, COR15am did not have a cryoprotective effect on isolated thylakoids.

In 1996, Uemura *et al.* (1996a) reported that both COR6.6 and COR15am invariably decreased the incidence of freeze-induced fusion of liposomes frozen *in vitro*. However, this effect, which was observed in liposomes composed of either a single species of phosphatidylcholine or a more complex mixture consisting of dioleoylphosphatidylethanolamine (DOPE) and dioleoylphosphatidylcholine (DOPC) and sterols or the lipid extract of the plasma membrane of rye leaves, only occurred under unusual conditions – when the liposomes were frozen in the absence of any solutes. Quite unexpectedly, the decreased incidence of freeze-induced fusion that was elicited by the COR polypeptides under these conditions coincided with an anomalous increase in freeze-induced leakage. Under no conditions tested was there a cryoprotective effect on freeze-induced leakage from the liposomes.

Webb *et al.* (1996) found that unlike sucrose, a well-known cryoprotectant, neither COR6.6 nor COR15am had an effect on the dehydration-induced increase in the liquid crystalline-to-gel phase transition temperature of either DPPC or DOPC. Similarly, neither COR6.6 nor COR15am altered the osmotic pressure (hydration) at which multilamellar vesicles composed of DOPE:DOPC underwent the lamellar-to-hexagonal II phase transition. However, when the vesicles were dehydrated in the presence of COR15am at osmotic pressures greater than 39 MPa, the resultant lipid aggregates had a polyhedral shape and the surface of the lamellae had a distinctive striated appearance. These observations suggested that COR15am interacts with lipid bilayers. However, collectively, these studies provided little evidence that either COR6.6 or COR15am has a direct cryoprotective effect on liposomes frozen *in vitro*. There are, however, many limitations inherent to *in vitro* freezing studies of liposomes (as exemplified by the studies of Uemura *et al.*, 1996a) and protein suspensions, and the results must be viewed very carefully before extrapolating to the freeze-induced destabilization of biological membranes frozen *in situ*.

A different approach to the quest to determine the role of the *COR* genes and polypeptides in the cold acclimation process became available when Thomashow and his colleagues were able to construct transgenic lines of *A. thaliana* that constitutively express the *COR* genes. When coupled with our characterization of specific freeze-induced lesions in *A. thaliana* (Uemura *et al.*, 1995), we were able to provide the first direct experimental evidence that at least one of the *COR* genes is functionally involved in the cold acclimation process (Artus *et al.*, 1996). Moreover, the ability to determine the effect of a single *COR* gene on specific lesions rather than merely survival *per se* has allowed for a more focused approach to the elucidation of their mode of action and the specific mechanism by which the COR polypeptides contribute to an increase in freezing tolerance.

2. Effect of *COR15a* on freezing tolerance

Using transgenic lines of *A. thaliana* that constitutively express the *COR15a* gene under non-acclimating conditions, we discovered that this gene affects the freezing tolerance of both chloroplasts frozen *in vivo* and protoplasts frozen *in situ* (Artus *et al.*, 1996). Nevertheless, the protoplast survival results were somewhat perplexing. Although constitutive expression of the *COR15a* gene resulted in a significant increase in protoplast survival over the range of −5 to −7°C, there was also a small, but very reproducible, decrease in survival over the range of −2 to −4°C. Furthermore, the mechanism by which the *COR15a* gene increased freezing tolerance was puzzling because the increase in survival was a manifestation of an increase in the cryostability of the plasma membrane and the COR15am polypeptide is reported to be located within the chloroplast stroma (Lin and Thomashow, 1992a).

In our most recent studies (Steponkus *et al.*, 1998), we have been able to (i) provide an explanation for both the positive and negative effects on protoplast survival, (ii) elucidate the mode of action of the *COR15a* gene, and (iii) propose and experimentally test a working hypothesis for the mechanism by which the COR15am polypeptide increases the cryostability of cellular membranes and freezing tolerance.

A comparison of the 'survival signatures' (survival vs freezing temperature) of protoplasts isolated from leaves of the transgenic *COR15a* plants (T8 line) and the wild-type RLD illustrates the negative and positive effects that expression of the *COR15a* gene has on survival (*Figure 1a*). Over the range of −2 to −4°C, survival of T8 protoplasts was less than that of the RLD; however, over the range of −5 to −7°C, survival of the T8 protoplasts was significantly greater. The apparent paradoxical effect of the *COR15a* gene on protoplast survival can be explained by consideration of the specific freeze-induced lesions that limit survival (see Steponkus *et al.*, 1993).

Over the range of −2 to −4°C, the predominant form of injury is expansion-induced lysis (EIL), which is a consequence of the osmotic excursions incurred during a freeze/thaw cycle (see Steponkus, 1991, for a review). At temperatures below −4°C, injury is manifested as a loss of osmotic responsiveness that is the result of freeze-induced lamellar-to-hexagonal II phase transitions (LOR-H_{II}) that involve the plasma membrane and various endomembranes − most often the chloroplast envelope − in regions where they are brought into close apposition as a result of freeze-induced removal of water from the surfaces of membranes (see Steponkus *et al.*, 1995). Therefore, we hypothesized that the lower survival of T8 protoplasts over the range of −2 to −4°C was a result of an increase in the incidence of EIL and the higher survival over the range of −5 to −7°C was the result of a decrease in the incidence of LOR-H_{II}. Both of these predictions were confirmed experimentally.

EIL can be precluded by subjecting the protoplasts to a freeze/hypertonic thaw

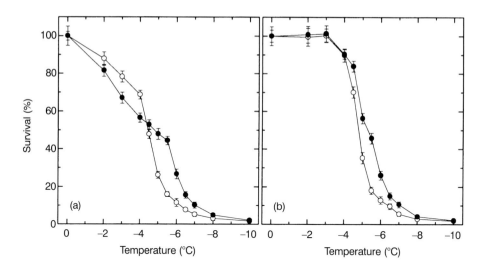

Figure 1. Freezing tolerance of protoplasts isolated from leaves of non-acclimated plants of *A. thaliana*: RLD (○), the wild type, and T8 (●), a transgenic line that constitutively expresses the *COR15a* gene. Protoplasts were suspended in a 0.400 osmolal sorbitol solution before freezing. (a) Survival after a conventional freeze/thaw treatment; (b) Survival after a freeze/hypertonic thaw treatment. Survival was determined by staining with fluorescein diacetate. Results shown are the mean and SD of three experiments. Reprinted from Steponkus, P.L. *et al.* Mode of action of the *COR15a* gene on the freezing tolerance of *Arabidopsis thaliana*. *Proceedings of the National Academy of Sciences, U.S.A.*, vol. 95, pp. 14570–14575. Copyright 1998 National Academy of Sciences, U.S.A.

treatment to limit the extent of osmotic expansion after thawing of the suspending medium (see Uemura and Steponkus, 1989). Whereas survival of protoplasts subjected to a conventional freeze/thaw cycle reflects the combined incidence of EIL and LOR-H_{II}, survival of protoplasts subjected to a freeze/hypertonic thaw treatment is limited only by LOR-H_{II} (*Figure 1b*). The incidence of EIL can be calculated as the difference in survival between the two treatments.

Such studies demonstrated that the incidence of EIL was indeed higher in T8 protoplasts. Although the reason for this was not immediately evident, we subsequently discovered that there is a small difference in the intracellular osmolality of T8 and RLD protoplasts. In our initial studies, we determined that 0.400 osmolal (osm) was the optimum tonicity (slightly hypertonic) for isolating protoplasts from RLD leaves. *A priori*, we assumed that the same osmolality would be appropriate for the isolation of protoplasts from T8 leaves. However, this was not a correct assumption. Leaf plasmolysis tests revealed that the intracellular tonicity was slightly higher in T8 leaves and that a 0.413 osm solution was the optimum tonicity.

In subsequent experiments in which both the T8 and RLD protoplasts were isolated and frozen in either a 0.413 or 0.425 osm sorbitol medium, the incidence of EIL was the same in both T8 and RLD protoplasts (*Figure 2*). Thus, the negative effect (i.e. an increased incidence of EIL) that was observed with T8 protoplasts in our initial study (Artus *et al.*, 1996) when the protoplasts were suspended in a 0.400 osm solution, was

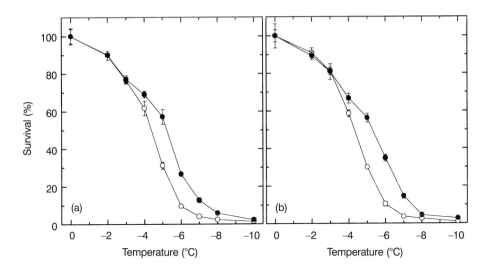

Figure 2. Freezing tolerance of protoplasts isolated from leaves of non-acclimated plants of *A. thaliana*: RLD (○) and T8 (●). Protoplasts were suspended in either a (a) 0.413 or (b) 0.425 osmolal sorbitol solution before freezing. Survival was determined by staining with fluorescein diacetate. Results shown are the mean and SD of three experiments. Part (a) reprinted from Steponkus, P.L. *et al.* Mode of action of the *COR15a* gene on the freezing tolerance of *Arabidopsis thaliana*. *Proceedings of the National Academy of Sciences, U.S.A.*, vol. 95, pp. 14570–14575. Copyright 1998 National Academy of Sciences, U.S.A.

in fact an artifact due to the T8 protoplasts not being suspended in a solution of the optimum tonicity. Because the T8 protoplasts have a higher internal osmolality (a steeper slope in a Boyle van't Hoff plot of volume vs osm⁻¹), they will undergo larger surface area changes during freeze/thaw-induced osmotic contraction/expansion and have an increased incidence of EIL (see Steponkus, 1991). The reason for the small difference in the intracellular osmolality (0.013) is not known at this time.

3.   Mode of action of the *COR15a* gene

A comparison of the survival signatures of RLD and T8 protoplasts after a freeze/hypertonic thaw treatment when suspended in a 0.400 osm solution (*Figure 1b*) or after a conventional freeze/thaw treatment when suspended in either a 0.413 or a 0.425 osm solution (*Figure 2*) demonstrates that expression of the *COR15a* gene results in an increase in protoplasts survival over the range of –4.5 to –7°C. As a result, the survival curve is shifted by ~1°C. The fact that the increase in survival occurred only over a limited range is not unexpected. With RLD protoplasts, the incidence of LOR-H$_{II}$ increases from <10% to >90% between –4 and –6°C.

The increase in survival elicited by expression of the *COR15a* gene is not uniquely observed in isolated protoplasts. An increase in freezing tolerance has also been observed with excised leaves (in which freezing injury is measured by electrolyte leakage). The increased freezing tolerance of T8 leaves occurs over the same approximate

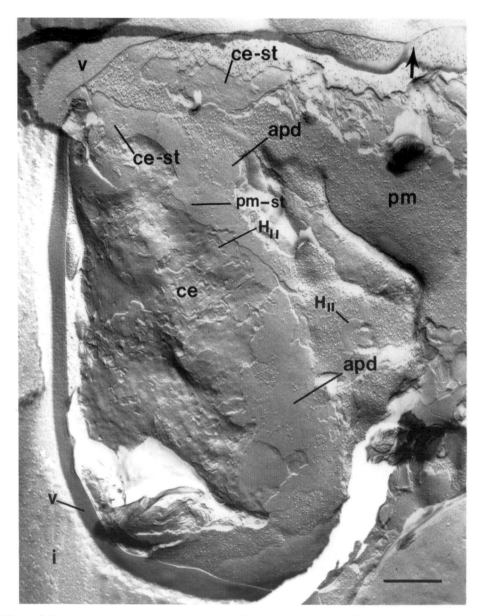

Figure 3. Freeze-fracture electron micrograph of a T8 protoplast suspended in a 0.400 osmolal sorbitol solution and frozen to −6°C for 90 min. Protoplasmic fracture face of the plasma membrane (pm) showing a typical lamellar region, which is characterized by a random distribution of intra-membraneous particles (imp), melding into a region where the H_{II} phase has formed. The plasma membrane, which contains numerous aparticulate domains (apd), is overlaying the chloroplast envelope (ce) with the H_{II} phase appearing in localized domains. Well-ordered striations appear in both the plasma membrane (pm-st) and the chloroplast envelope (ce-st). Magnification: 40 100×; bar represents 0.5 nm; arrow indicates direction of shadowing. Modified from Steponkus, P.L. *et al.* Mode of action of the *COR15a* gene on the freezing tolerance of *Arabidopsis thaliana. Proceedings of the National Academy of Sciences, U.S.A.*, vol. 95, pp. 14570–14575. Copyright 1998 National Academy of Sciences, U.S.A.

temperature range (–3 to –7°C) and is of the same magnitude (~1°C) as that observed with isolated protoplasts (Uemura and Steponkus, 1998).

That the increased survival was a consequence of a decreased incidence of LOR-H_{II} was directly confirmed by freeze-fracture electron microscopy (FFEM) studies. Formation of the H_{II} phase, which is a three-dimensional array of inverted cylindrical micelles, is an inter-bilayer event that involves two or more membranes that are brought into close apposition as a result of freeze-induced dehydration (Steponkus *et al.*, 1993, 1995). The H_{II} phase is observed most often in regions where the plasma membrane is in close apposition with the chloroplast envelope, which is comprised of two membranes – the inner and outer membrane (*Figure 3*). In addition to the tightly packed cylinders characteristic of the H_{II} phase, two other ultrastructural morphologies are commonly observed to be associated with freeze-induced formation of the H_{II} phase: well-ordered striations and loosely ordered swirls in regions of the plasma membrane and chloroplast envelope (*Figure 4*). These morphologies are thought to be intermediates in the lamellar-to-hexagonal II phase transition.

The incidence of the H_{II} phase and the associated morphologies determined in the FFEM studies closely paralleled protoplast mortality (% mortality = 100% – % survival) over the range of –4.5 to –7°C. For instance, at –5.5°C, which is the temperature

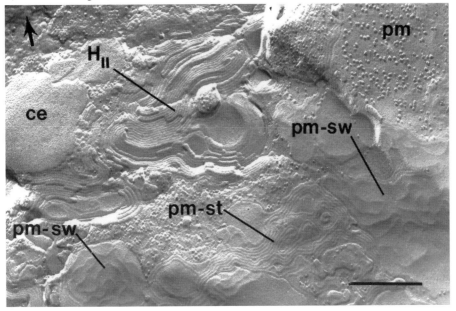

Figure 4. Freeze-fracture electron micrograph of an RLD protoplast suspended in a 0.400 osmolal sorbitol solution and frozen to –6°C for 90 min. High magnification (78 500×) micrograph illustrating the tightly packed cylinders that are characteristic of the H_{II} phase and two other ultrastructural morphologies commonly observed in association with freeze-induced formation of the H_{II} phase in biological cells: well-ordered striations (pm-st) and loosely ordered swirls (pm-sw) in the plasma membrane. Bar represents 0.25 nm; arrow indicates direction of shadowing. Modified from Steponkus, P.L. *et al.* Mode of action of the *COR15a* gene on the freezing tolerance of *Arabidopsis thaliana*. *Proceedings of the National Academy of Sciences, U.S.A.*, vol. 95, pp. 14570–14575. Copyright 1998 National Academy of Sciences, U.S.A.

at which there was the maximum difference in mortality of RLD (83%) and T8 (56%) protoplasts, the various H_{II} morphologies were observed in 76% (82/108) of the RLD protoplasts and 57% (100/176) of the T8 protoplasts. At $-5°C$, there was 65% mortality of RLD protoplasts and the H_{II} phase was observed in 63% of the protoplasts; in T8, there was 44% mortality and the H_{II} phase was observed in 37% of the protoplasts. Like the survival signatures, the incidence of freeze-induced formation of the H_{II} phase determined in the FFEM studies was shifted by ~1°C.

4. Mechanism by which COR15am increases freezing tolerance

Although these studies establish that expression of the *COR15a* gene decreases the propensity for freeze-induced formation of the H_{II} phase (LOR-H_{II}), the results were somewhat puzzling and prompted an obvious question. If COR15am is located in the chloroplast stroma, how can it decrease the incidence of freeze-induced formation of the H_{II} phase, which is an inter-bilayer phenomenon that most often involves the inner and outer membrane of the chloroplast envelope and the plasma membrane?

As a working hypothesis, we have proposed (Steponkus *et al.*, 1998) that (i) the onset (freezing temperature) of the H_{II} phase is determined by the membrane in the ensemble consisting of the inner and outer membranes of the chloroplast envelope and the plasma membrane that has the greatest propensity to form the H_{II} phase, that is the 'weak link' in the ensemble, (ii) that the inner membrane of the chloroplast envelope has the greatest propensity to form the H_{II} phase, and (iii) the COR15am polypeptide, which is predicted to be composed of amphipathic α-helical regions, alters the intrinsic curvature of the monolayers that comprise the inner membrane of the chloroplast envelope.

The supposition that the inner membrane of the chloroplast envelope is the 'weak link' in the ensemble that undergoes the freeze-induced formation of the H_{II} phase is based on analyses of the lipid composition of the inner and outer membranes of the chloroplast envelope of winter rye (Uemura and Steponkus, 1997). The inner membrane contains 48 mol% monogalactosyldiacylglycerol (MGDG), which has a very high propensity to form the H_{II} phase – the outer membrane contains only 20 mol%. The outer membrane contains 30 mol% phosphatidylcholine (PC), which stabilizes the bilayer configuration – the inner membrane contains only 8 mol%. Both contain a similar proportion of digalactosyldiacylglycerol (DGDG) (30 mol%) and sulfoquinovosyldiacylglycerol (SQDG) (7 mol%), which also stabilize the bilayer configuration. ^{31}P-NMR studies of the phase behaviour of lipid mixtures composed of the major lipid classes that comprise the chloroplast envelope (MGDG:DGDG:SQDG:PC) in proportions that are similar to those found in the inner and outer membrane of the chloroplast envelope have provided direct evidence that the inner membrane mixture has a greater propensity to form non-lamellar phases than does the mixture in proportions similar to the outer membrane (Uemura *et al.*, 1996b).

The hypothesis that the COR15am polypeptide alters the intrinsic curvature of the inner membrane of the chloroplast envelope is based on reports that amphipathic polypeptides that form an α-helix can have a strong effect on the intrinsic curvature of monolayers that comprise a lipid bilayer and that a shift in the lamellar-to-hexagonal II phase transition temperature (T_{bh}) is a very sensitive indicator of the effect on monolayer curvature (Epand *et al.*, 1995). Indeed, ^{31}P-NMR studies revealed that COR15am increased the T_{bh} of DOPE vesicles by ~2°C, which is the same magnitude as the shift in the survival curve of isolated protoplasts and the shift in the incidence of the H_{II} phase as determined in the FFEM studies. These results were confirmed in X-ray diffraction studies, which also revealed that COR15am increased the hexagonal lattice spacing of the H_{II} phase of DOPE between 5 and 15°C.

In subsequent studies, we observed that COR15am also affects the phase behaviour of a complex mixture composed of the major lipid species in the chloroplast envelope (MGDG:DGDG:SQDG:PC:PG (phosphatidylglycerol)) in proportions that resemble the inner membrane (50:30:5:10:5 mol%). In the absence of the COR15am polypeptide, the ^{31}P-NMR spectrum appeared to be a composite of several phases (lamellar, non-lamellar and possibly cubic), which is commonly reported for mixtures of MGDG:DGDG and lipid extracts of chloroplasts. With lipid mixtures containing the COR15am polypeptide, the ^{31}P-NMR spectrum at 0°C was characteristic of the lamellar phase.

5. *COR* genes in perspective

These studies have established the specific role of the *COR15a* gene in the cold acclimation process: to decrease the propensity for freeze-induced formation of the H_{II} phase (LOR-H_{II}) – the lesion that limits the freezing tolerance of non-acclimated leaves of *A. thaliana* and winter cereals, such as rye, wheat, barley and oat. In these species, the propensity for freeze-induced formation of the H_{II} phase is decreased during the initial stages of cold acclimation such that it does not occur in cold-acclimated leaves. In *A. thaliana*, freeze-induced formation of the H_{II} phase is precluded after just 2–3 days of cold acclimation. In winter rye, it is precluded after the first week of cold acclimation. In both, the maximum increase in freezing tolerance is not attained until after 4 or 5 weeks of cold acclimation.

Although constitutive expression of the *COR15a* gene decreases the incidence of freeze-induced formation of the H_{II} phase, it does not preclude its occurrence – as occurs in the initial stages of cold acclimation. Therefore, other factors such as low-temperature induced membrane lipid alterations (Steponkus *et al.*, 1993) and the accumulation of sugars (Steponkus and Uemura, 1998a), which also occur during the early stages of cold acclimation, are also involved in decreasing the propensity for freeze-induced formation of the H_{II} phase. Preliminary studies in which the intracellular sugar content can be artificially manipulated suggest that the accumulation of sugars (predominantly glucose, fructose and sucrose in *A. thaliana*) and alterations in membrane lipid composition have a greater effect on minimizing freeze-induced formation of the H_{II} phase than does *COR15a*. It is important to note that after just one day of cold acclimation, the freezing tolerance of the RLD protoplasts is similar to that of the T8 protoplasts; this is also observed with excised leaves (M. Uemura and P.L. Steponkus, unpublished data).

Nevertheless, there are several other *COR* genes (e.g. *COR6.6*, *COR47* and *COR78*), and it is not unreasonable to expect that they may act collectively to increase freezing tolerance to a greater extent. Indeed, we have observed an increase (~1°C) in the freezing tolerance of excised leaves of transgenic lines that constitutively express the *COR6.6* gene (Uemura and Steponkus, 1998). Moreover, the effect of the *COR6.6* gene appears to be additive to that of the *COR15a* gene in that the freezing tolerance of leaves from a transgenic line that expresses both the *COR15a* and *COR6.6* genes is greater (~1°C) than that of leaves from lines that express either one alone.

Recently, Thomashow's group reported that constitutive over-expression of the *Arabidopsis* transcriptional activator CBF1 induces expression of CRT/DRE-regulated cold-responsive genes and results in about a 3°C increase in the freezing tolerance of *A. thaliana* (Jaglo-Ottosen *et al.*, 1998). At the same time, Shinozaki's group reported that over-expression of the DREB1A protein, which binds to the *cis*-acting drought-responsive element, also increased the freezing tolerance of *A. thaliana* plants (Liu *et al.*, 1998). In the report of Jaglo-Ottosen *et al.* (1998), the freezing tolerance (expressed as the EL_{50}, i.e. the freezing temperature that results in 50% electrolyte leakage from excised leaves) of the transgenic A6 line (–7.2°C) was similar to that (–7.6°C) determined for the wild type (RLD) after 7–10 days of cold acclimation. Although this might be construed to mean that the expression of the full complement of the *COR* genes is responsible for a large portion of the increase in freezing tolerance that can be achieved, we do not believe that this is a correct inference. The values given for the cold-acclimated wild type are substantially less than the maximum freezing tolerance that is attainable for this ecotype (RLD). Whereas Jaglo-Ottosen *et al.* report an EL_{50} value of –7.6°C for leaves of RLD plants that were cold acclimated for 7–10 days, we consistently observe that the EL_{50} is –9.9°C after 7 days of cold acclimation and –10.5°C after 10 days of acclimation. The maximum freezing tolerance that we have achieved with RLD is –12.6°C after 35 days of cold acclimation, which is nearly 8°C lower than the EL_{50} that we determined for non-acclimated RLD leaves (Steponkus and Uemura, 1998b). Clearly, the *COR* genes contribute to an increase in freezing tolerance; however – assuming that the plants used in the studies of Jaglo-Ottosen *et al.* (1998) are expressing the full complement of the *COR* genes at a high level – expression of the *COR* genes does not account for the majority of the maximum freezing tolerance that is attainable.

Cognizance of this fact is of considerable importance for future directions of research in this field and the potential use of the *COR* genes and/or their promoters for increasing the freezing tolerance of crop species. The cold acclimation process is an amalgam of many different low temperature-induced changes. Expression of the *COR* genes is but one facet of this complex developmental process. Other metabolic changes that are induced by low temperature, most notably alterations in membrane lipid composition and the accumulation of endogenous cryoprotectants, such as sucrose and other sugars, play an equally important role. These three factors – alterations in membrane lipid composition, the accumulation of sugars and other cryoprotective solutes, and the expression of the *COR* genes – appear to act additively in increasing freezing tolerance of *A. thaliana* (Steponkus and Uemura, 1998a) and may be considered as the trinity of cold acclimation.

6. Future challenges

The challenge that lies ahead is to determine the specific roles of the other *COR* genes and to elucidate the relative contributions of low temperature-induced alterations in membrane lipid composition and sugar accumulation, which appear more likely to limit the maximum freezing tolerance that is attainable. As such, they – rather than the *COR* genes – may provide the means by which an increase in the maximum freezing tolerance of crop species is achieved through genetic engineering approaches.

Acknowledgements

These studies were supported by grants from the U.S. Department of Agriculture National Research Initiative Competitive Grants Program (96–35100–3163) and the U.S. Department of Energy (DE-FG01–84ER13214). Seeds of the transgenic lines that constitutively produce the COR polypeptides were provided by M.F. Thomashow and S.J. Gilmour.

References

Artus, N.N., Uemura, M., Steponkus, P.L., Gilmour, S.J., Lin, C. and Thomashow, M.F. (1996) Constitutive expression of the cold-regulated *Arabidopsis thaliana COR15a* gene affects chloroplast and protoplast freezing tolerance. *Proc. Natl Acad. Sci. USA* **93:** 13404–13409.

Cattivelli, L. and Bartels, D. (1990) Molecular cloning and characterization of cold-regulated genes in barley. *Plant Physiol.* **93:** 1504–1510.

Crosatti, C., Nevo, E., Stanca, A.M. and Cattivelli, L. (1996) Genetic analysis of the accumulation of COR14 proteins in wild (*Hordeum spontaneum*) and cultivated (*Hordeum vulgare*) barley. *Theor. Appl. Genet.* **93:** 975–981.

Epand, R.M., Shai, Y., Segrest, J.P. and Anantharamaiah, G.M. (1995) Mechanisms for the modulation of membrane bilayer properties by amphipathic helical peptides. *Biopolymers* **37:** 319–338.

Gilmour, S.J., Artus, N.N. and Thomashow, M.F. (1992) cDNA sequence analysis and expression of two cold-regulated genes of *Arabidopsis thaliana. Plant Mol. Biol.* **18:** 13–22.

Guy, C.L. and Haskell, D.L. (1987) Induction of freezing tolerance in spinach is associated with the synthesis of cold acclimation induced proteins. *Plant Physiol.* **84:** 872–878.

Guy, C.L., Niemi, K.J. and Brambl, R. (1985) Altered gene expression during cold acclimation of spinach. *Proc. Natl Acad. Sci. USA* **82:** 3673–3677.

Hincha, D.K., Sieg, F., Bakaltcheva, I., Köth, H. and Schmitt, J.M. (1996). Freeze-thaw damage to thylakoid membranes: specific protection by sugars and proteins. In: *Advances in Low-Temperature Biology,* Vol. 3 (ed. P.L. Steponkus). JAI Press, London, pp. 141–183.

Jaglo-Ottosen, K.R., Gilmour, S.J., Zarka, D.G., Schabenberger, O. and Thomashow, M.F. (1998) *Arabidopsis CBF1* overexpression induces COR genes and enhances freezing tolerance. *Science* **280:** 104–106.

Leyva, A., Jarillo, J.A., Salinas, J. and Martinez-Zapater, J.M. (1995) Low temperature induces the accumulation of *phenylalanine ammonia-lyase* and *chalcone synthase* mRNAs of *Arabidopsis thaliana* in a light-dependent manner. *Plant Physiol.* **108:** 39–46.

Lin, C. and Thomashow, M.F. (1992a) DNA sequence analysis of a complementary DNA for cold-regulated *Arabidopsis* gene *cor15* and characterization of the COR15 polypeptide. *Plant Physiol.* **99:** 519–525.

Lin, C. and Thomashow, M.F. (1992b) A cold-regulated *Arabidopsis* gene encodes a polypeptide having potent cryoprotective activity. *Biochem. Biophys. Res. Commun.* **183:** 1103–1108.

Liu, Q., Kasuga, M., Sakurai, Y., Abe, H., Miura, S., Yamaguchi-Shinozaki, K. and Shinozaki, K. (1998) Two transcription factors, DREB1 and DREB2, with an EREBP/AP2 DNA binding domain separate two cellular signal transduction pathways in drought- and low-temperature-responsive gene expression, respectively, in Arabidopsis. *Plant Cell* **10:** 1391–1406.

Luo, M., Liu, J.-H., Mohapatra, S., Hill, R.D. and Mohapatra, S.S. (1992) Characterization of a gene family encoding abscisic acid- and environmental stress-inducible proteins of alfalfa. *J. Biol. Chem.* **267:** 15367–15374.

Mohapatra, S.S., Poole, R.J. and Dhindsa, R.S. (1987) Cold acclimation, freezing resistance and protein synthesis in alfalfa (*Medicago sativa* L. cv. Saranac). *J. Exp. Bot.* **38**: 1697–1703.

Mohapatra, S.S., Wolfraim, L., Poole, R.J. and Dhindsa, R.S. (1989) Molecular cloning and relationship to freezing tolerance of cold-acclimation-specific genes of alfalfa. *Plant Physiol.* **89**: 375–380.

Nordin, K., Vahala, T. and Palva, E.T. (1993) Differential expression of two related, low-temperature-induced genes in *Arabidopsis thaliana* (L.) Heynh. *Plant Mol. Biol.* **21**: 641–653.

Steponkus, P.L. (1991) Behavior of the plasma membrane during osmotic excursions. In: *Endocytosis, Exocytosis, and Vesicle Traffic in Plants* (eds C.R. Hawes, J.O.D. Coleman and D.E. Evans). SEB Sem. Series: 45. Cambridge University Press, Cambridge, pp. 103–128.

Steponkus, P.L. and Uemura, M. (1998a) Resolution of the individual contributions of membrane lipid alterations, sugar accumulation, and the *COR15a* gene to the freezing tolerance of *Arabidopsis thaliana*. *Cryobiology* **37**: 391–392.

Steponkus, P.L. and Uemura, M. (1998b) Freezing tolerance of *Arabidopsis thaliana* mutants that constitutively express the *COR15a* and *CBF1* genes. 9th International Conference on Arabidopsis Research, Abstract #143.

Steponkus, P.L., Uemura, M. and Webb, M.S. (1993) A contrast of the cryostability of the plasma membrane of winter rye and spring oat – two species that widely differ in their freezing tolerance and plasma membrane lipid composition. In: *Advances in Low-Temperature Biology,* Vol. 2 (ed. P.L. Steponkus). JAI Press, London, pp. 211–312.

Steponkus, P.L., Uemura, M. and Webb, M.S. (1995) Freeze-induced destabilization of cellular membranes and lipid bilayers. In: *Permeability and Stability of Lipid Bilayers* (ed. E.A. Disalvo and S.A. Simon). CRC Press, Boca Raton, FL, pp. 77–104.

Steponkus, P.L., Uemura, M., Joseph, R.A., Gilmour, S.J. and Thomashow, M.F. (1998) Mode of action of the *COR15a* gene on the freezing tolerance of *Arabidopsis thaliana*. *Proc. Natl Acad. Sci. USA* **95**:14570–14575.

Thomashow, M.F. (1990) Molecular genetics of cold acclimation in higher plants. *Adv. Genet.* **28**: 99–131.

Thomashow, M.F. (1993) Genes induced during cold acclimation in higher plants. In: *Advances in Low-Temperature Biology,* Vol. 2 (ed. P.L. Steponkus). JAI Press, London, pp. 183–210.

Thomashow, M.F., Gilmour, S.J., Hajela, R., Horvath, D., Lin, C. and Guo, W. (1990) Studies on cold acclimation in *Arabidopsis thaliana*. In: *Horticultural Biotechnology* (ed. A.B. Bennett and S.D. O'Neill). Wiley-Liss, New York, pp. 305–314.

Uemura, M. and Steponkus, P.L. (1989) Effect of cold acclimation on the incidence of two forms of freezing injury in protoplasts isolated from rye leaves. *Plant Physiol.* **91**: 1131–1137.

Uemura, M. and Steponkus, P.L. (1997) Effect of cold acclimation on the lipid composition of the inner and outer membrane of the chloroplast envelope isolated from rye leaves. *Plant Physiol.* **114**: 1493–1500.

Uemura, M. and Steponkus, P.L. (1998) Freezing tolerance of *Arabidopsis thaliana* mutants that constitutively express the *COR15a, COR6.6* and *CBF1* genes. *Cryobiology* **37**: 390–391.

Uemura, M., Gilmour, S.J., Thomashow, M.F. and Steponkus, P.L. (1994) Effect of COR proteins on the cryostability of chloroplast thylakoids isolated from spinach leaves. *Cryobiology* **31**: 558.

Uemura, M., Joseph, R.A. and Steponkus, P.L. (1995) Cold acclimation of *Arabidopsis thaliana* – effect on plasma membrane lipid composition and freeze-induced lesions. *Plant Physiol.* **109**: 15–30.

Uemura, M., Gilmour, S.J., Thomashow, M.F. and Steponkus, P.L. (1996a) Effects of COR6.6 and COR15am polypeptides encoded by *COR* (Cold-Regulated) genes of *Arabidopsis thaliana* on the freeze-induced fusion and leakage of liposomes. *Plant Physiol.* **111**: 313–327.

Uemura, M., Gilmour, S.J., Thomashow, M.F. and Steponkus, P.L. (1996b) Effect of COR15am on the lamellar-to-hexagonal II phase transition temperature of phospholipids and lipid mixtures. *Cryobiology* **33**: 657.

Volger, H.G. and Heber, U. (1975) Cryoprotective leaf proteins. *Biochim. Biophys. Acta* **412**: 335–349.

Webb, M.S., Gilmour, S.J., Thomashow, M.F. and Steponkus, P.L. (1996) Effects of COR6.6 and COR15am polypeptides encoded by *COR* (Cold-Regulated genes of *Arabidopsis thaliana* on dehydration-induced phase transitions of phospholipid membranes. *Plant Physiol.* **111**: 301–312.

Chapter 14

Cryoprotection of thylakoid membranes by soluble plant proteins

Dirk K. Hincha, Gilbert Tischendorf, Birgit Neukamm and Jürgen M. Schmitt

1. Background

There is a great variability in the frost hardiness (i.e. the temperature at which cellular damage occurs during freezing) of different plant species. In addition, many plants are able to increase their frost hardiness during a cold acclimation period at a low, but non-freezing temperature (typically between 0 and 10°C; see Hincha and Schmitt, 1992b; Levitt, 1980; Steponkus, 1984 for reviews). Over the last few years, a large array of cold-regulated (*COR*) genes has been cloned from different species (see Hughes and Dunn, 1996 for a recent review). The proteins encoded by these genes are thought to play an important role in plant frost hardiness. Key questions in this area at the moment include:

(i)   Which of these proteins have functional significance for frost hardiness?
(ii) What is the quantitative contribution of the different proteins to plant frost hardiness?
(iii) Which of these proteins have cryoprotective activity (i.e. are directly able to stabilize cellular structures during a freeze–thaw cycle), and which act indirectly through their enzymatic activity and/or as signal transduction molecules?
(iv) What functional mechanisms are employed by cryoprotective proteins and how is their function related to their structure?

To establish the cryoprotective activity of COR proteins *in vivo,* expression in transgenic plants is employed. This has been done by several investigators, but only a small increase in frost hardiness has been reported so far (Artus *et al.,* 1996). It seems likely that the co-ordinate expression of more than one cold-induced gene is necessary to achieve larger increases in frost hardiness.

To test potentially cryoprotective proteins *in vitro,* a meaningful assay system has to be established. This requires knowledge about the mechanisms leading to freeze–thaw damage and how damage is ameliorated during hardening. This has been established in detail only for the plasma membrane (Steponkus, 1984) and chloroplast thylakoids (Hincha *et al.,* 1996).

Plant Responses to Environmental Stress, edited by M.F. Smallwood, C.M. Calvert and D.J. Bowles.
© 1999 BIOS Scientific Publishers, Oxford.

2. Thylakoids as an assay system for the detection of cryoprotective proteins

Photosynthesis is a frost-sensitive physiological activity of plant leaves. A key factor in the frost-induced inactivation of photosynthesis is the inactivation of electron transport. In spinach (*Spinacia oleracea* L.), this could be related to the loss of the electron transport protein plastocyanin from the thylakoid lumen, which occurs when leaves are thawed after freezing to a lethal temperature (Hincha and Schmitt, 1992b). Loss of plastocyanin can also be induced during a freeze–thaw cycle *in vitro*. The release of plastocyanin from isolated thylakoids is the result of transient membrane rupture, which also leads to the loss of other internal proteins and solutes, and to a reduction in thylakoid volume (compare *Figure 1a* and *1b*). This rupture is, at least in part, the result of solute loading of the vesicles. During freezing, ice forms from pure water and the membranes are trapped together with the solutes in the unfrozen solution which remains in equilibrium with the ice. The high solute concentrations reached in such partially frozen systems drive the diffusion of solutes across the membranes. This leads to solute loading and to osmotic swelling of the thylakoids, and finally to rupture during thawing, when the ice melts and the solution surrounding the membranes is diluted back to the initial concentration. The activity of cryoprotective proteins can be quantitated either as a reduction in plastocyanin release, or as the conservation of thylakoid volume after a freeze–thaw cycle, measured as the packed volume of the membranes after haematocrit centrifugation (Hincha and Schmitt, 1992a).

3. Proteins with cryoprotective activity for thylakoids

3.1 β-1,3-Glucanase

Glucanases are enzymes that belong to the group of pathogenesis-related (PR) proteins. A class I β-1,3-glucanase from tobacco has cryoprotective activity and immunologically related proteins are accumulated in cabbage (*Brassica oleracea* L.) and spinach during frost hardening under natural conditions. In contrast to other COR proteins, they are not induced at 4°C, but probably require lower temperatures and/or freezing (Hincha *et al.*, 1997b).

3.2 Lectins

Lectins are sugar-binding proteins, which are classified according to their specificity for different monosaccharides. Since thylakoids contain galactolipids (Webb and Green, 1991), galactose-specific lectins bind to the membrane surface and some exert a cryoprotective effect (Hincha *et al.*, 1993, 1997c). Cryoprotection requires binding to digalactolipid headgroups and the cryoprotective efficiency of different lectins is linearly related to their relative hydrophobicity. It was found that, after binding to the galactolipid headgroups, hydrophobic interactions with membrane lipids lead to reduced fluidity at the membrane surface, resulting in reduced solute permeability (Hincha *et al.*, 1997a). This reduces solute diffusion during freezing and therefore

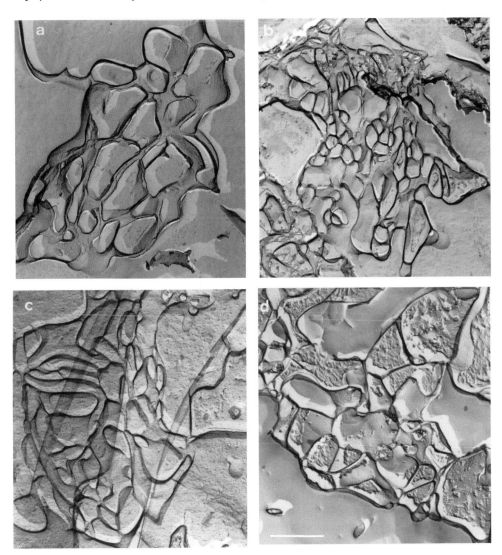

Figure 1. Freeze-fracture electron micrographs of thylakoids isolated from non-acclimated spinach plants. All samples contained 2.5 mM NaCl, 5 mM sucrose, and thylakoids corresponding to approximately 0.5 mg chlorophyll ml^{-1}. In addition, samples in (c) and (d) contained cryoprotectin at a concentration resulting in complete cryoprotection in a haematocrit centrifugation assay (see text). Samples in (a) and (c) were controls stored for 1 h at 0°C, while samples in (b) and (d) were frozen for 1 h at −20°C and then thawed. For freeze-fracture replication, all samples were frozen in liquid freon 22 cooled with liquid N_2 and subsequently transferred to liquid N_2. They were fractured using a Balzers BA 360M freeze-fracture apparatus. Freeze-etched fracture faces were coated with platinum and carbon. Replicas were cleaned with 50% chloroform/50% methanol (v/v) and were examined in a Siemens 101 transmission electron microscope. The scale bar in (d) represents 1 μm. Magnification was the same in all four panels.

osmotic membrane rupture during thawing, which was measured as reduced plasto-cyanin release. Most lectins that have been assayed are commercially available seed lectins (Hincha *et al*., 1993), which play no role in plant frost hardiness. However, two cryoprotective lectins from mistletoe (*Viscum album* L.) leaves showed increased concentrations during the winter months, when leaf frost hardiness is high, and low concentrations in the summer when the leaves are susceptible to freezing damage (Hincha *et al*., 1997c). Another mistletoe leaf lectin, that is not cryoprotective *in vitro*, shows no seasonal variation in concentration.

3.3 Cryoprotectin

This protein has been purified from leaves of cold-acclimated cabbage plants (Sieg *et al*., 1996). It is a 7-kDa protein that is stable to boiling (Hincha and Schmitt, 1992a), low pH and acetonitrile (Sieg *et al*., 1996), and 6 M urea (B. Neukamm and D.K. Hincha, unpublished results). Partial sequencing showed that it is a member of the class of non-specific lipid transfer proteins (LTPs; D.K. Hincha, W. Schröder, F. Seig and J.M. Schmitt, unpublished data; see Kader, 1996 for a comprehensive review on plant LTPs). It reduces the loss of plastocyanin and preserves thylakoid volume during a freeze–thaw cycle (*Figure 1*; Hincha and Schmitt, 1992a). This cryoprotective activity is dependent on the presence of Ca and Mn, and can be completely inhibited by pretreatment of the protein with ethylenediamine tetraacetic acid (EDTA) (F. Sieg, J.M. Schmitt and D.K. Hincha, unpublished data). Recent evidence suggests that cryoprotectin binds to thylakoid membranes (A. Bulstra, B. Neukamm and D.K. Hincha, unpublished data). In contrast to the closely related protein WAX9 (Pyee *et al*., 1994), cryoprotectin has no lipid transfer activity. WAX9, like several other plant LTPs, has no cryoprotective activity (B. Neukamm, J.M. Schmitt and D.K. Hincha, unpublished data). Cryoprotectin is only found in the leaves of cold-acclimated plants (Hincha *et al*., 1990; Sieg *et al*., 1996), while WAX9 is present in non-acclimated plants. Cold-induced LTP genes have been described in barley (Hughes *et al*., 1992), but no functional characterization of the encoded proteins has been published.

4. Functional comparison

For all cryoprotective proteins described above, we have shown that they reduce solute loading during freezing. Whether they all do this by the same mechanism remains to be resolved. Cryoprotectin also seems to have additional stabilizing effects for thylakoids during freezing (Hincha *et al*., 1990). The relative cryoprotective efficiency (i.e. the concentration necessary to achieve a specific degree of protection) of the different proteins varies widely. The least effective lectin (mistletoe lectin ML II) and the most effective lectin (*Ricinus communis* lectin RCA_{60}) differ by a factor of approximately 20, and β-1,3-glucanase is about half as effective as RCA_{60}. In comparison, cryoprotectin is approximately 1000-fold more efficient than this lectin and approximately 10^6 times more efficient than sucrose when compared on a molar basis.

5. Future directions

The *in vivo* analysis of subsets of COR proteins will profit from the cloning and expression in transgenic plants of specific transcription factors (Stockinger *et al.*, 1997), which has already yielded first encouraging results (Jaglo-Ottosen *et al.*, 1998). To establish more *in vitro* models to test the cryoprotective activity of cold-induced proteins, knowledge of the cryobehaviour of additional cellular structures will be necessary. The mode of action of known cryoprotective proteins will have to be established through a combination of biophysical and biochemical methods, including X-ray crystallography and site-directed mutagenesis.

Acknowledgements

Supported by grants (to D.K.H. and J.M.S.) and a Heisenberg stipend (to D.K.H.) from the Deutsche Forschungsgemeinschaft.

References

Artus, N.N., Uemura, M., Steponkus, P.L., Gilmour, S.J., Lin, C. and Thomashow, M.F. (1996) Constitutive expression of the cold-regulated *Arabidopsis thaliana* COR15a gene affects both chloroplast and protoplast freezing tolerance. *Proc. Natl Acad. Sci. USA* **93**: 13404–13409.

Hincha, D.K. and Schmitt, J.M. (1992a) Cryoprotective leaf proteins: assay methods and heat stability. *J. Plant Physiol.* **140**: 236–240.

Hincha, D.K. and Schmitt, J.M. (1992b) Freeze–thaw injury and cryoprotection of thylakoid membranes. In: *Water and Life* (eds. G.N. Somero, C.B. Osmond and C.L. Bolis). Springer, Berlin, pp. 316–337.

Hincha, D.K., Heber, U. and Schmitt, J.M. (1990) Proteins from frost-hardy leaves protect thylakoids against mechanical freeze–thaw damage *in vitro*. *Planta* **180**: 416–419.

Hincha, D.K., Bakaltcheva, I. and Schmitt, J.M. (1993) Galactose-specific lectins protect isolated thylakoids against freeze–thaw damage. *Plant Physiol.* **103**: 59–65.

Hincha, D.K., Sieg, F., Bakaltcheva, I., Köth, H. and Schmitt, J.M. (1996) Freeze–thaw damage to thylakoid membranes: specific protection by sugars and proteins. In: *Advances in Low-temperature Biology*, Vol. 3 (ed. P.L. Steponkus). JAI Press, London, pp. 141–183.

Hincha, D.K., Bratt, P.J. and Williams, W.P. (1997a) A cryoprotective lectin reduces the solute permeability and lipid fluidity of thylakoid membranes. *Cryobiology* **34**: 193–199.

Hincha, D.K., Meins, F. Jr, Schmitt, J.M. (1997b) β-1,3-Glucanase is cryoprotective *in vitro* and is accumulated in leaves during cold acclimation. *Plant Physiol.* **114**: 1077–1083.

Hincha, D.K., Pfüller, U. and Schmitt, J.M. (1997c) The concentration of cryoprotective lectins in mistletoe (*Viscum album* L.) leaves is correlated with leaf frost hardiness. *Planta* **203**: 140–144.

Hughes, M.A. and Dunn, M.A. (1996) The molecular biology of plant acclimation to low temperature. *J. Exp. Bot.* **47**: 291–305.

Hughes, M.A., Dunn, M.A., Pearce, R.S., White, A.J. and Zhang, L. (1992) An abscisic-acid-responsive, low temperature barley gene has homology with a maize phospholipid transfer protein. *Plant Cell Environ.* **15**: 861–865.

Jaglo-Ottosen, K.R., Gilmour, S.J., Zarka, D.G., Schabenberger, O. and Thomashow, M.F. (1998) *Arabidopsis* CBF1 overexpression induces *COR* genes and enhances freezing tolerance. *Science* **280**: 104–106.

Kader, J.-C. (1996) Lipid-transfer proteins in plants. *Annu. Rev. Plant Physiol. Plant Mol. Biol.* **47**: 627–654.

Levitt, J. (1980) *Responses of Plants to Environmental Stresses, Vol. 1. Chilling, Freezing, and High Temperature Stresses.* Academic Press, Orlando, FL.

Pyee, J., Yu, H. and Kolattukudy, P.E. (1994) Identification of a lipid tranfer protein as the major protein in the surface wax of broccoli (*Brassica oleracea*) leaves. *Arch. Biochem. Biophys.* **311:** 460–468.

Sieg, F., Schröder, W., Schmitt, J.M. and Hincha, D.K. (1996) Purification and characterization of a cryoprotective protein (cryoprotectin) from the leaves of cold-acclimated cabbage. *Plant Physiol.* **111:** 215–221.

Steponkus, P.L. (1984) Role of the plasma membrane in freezing injury and cold acclimation. *Annu. Rev. Plant Physiol.* **35:** 543–584.

Stockinger, E.J., Gilmour, S.J. and Thomashow, M.F. (1997) *Arabidopsis thaliana* CBF1 encodes an AP2 domain-containing transcriptional activator that binds to the C-repeat/DRE, a *cis*-acting DNA regulatory element that stimulates transcription in response to low temperature and water deficit. *Proc. Natl. Acad. Sci. USA* **94**: 1035–1040.

Webb, M.S. and Green, B.R. (1991) Biochemical and biophysical properties of thylakoid acyl lipids. *Biochim. Biophys. Acta* **1060:** 133–158.

Chapter 15

Winter survival of transgenic *Medicago sativa* over-expressing superoxide dismutase

Bryan D. McKersie, Stephen R. Bowley, Kim S. Jones and Bruce Gossen

1. Background

Winter-hardiness is a complex trait involving tolerances to freezing, desiccation, ice-encasement (severe anoxia), flooding (milder anoxia) and disease. The combination and severity of these stresses that crops must tolerate varies with environment and year. Distinctly different physiological effects are caused by these different environmental stresses, but there may be at least one common element, oxidative stress. Freezing, anoxia and desiccation stresses have been linked with oxidative stress in three types of correlative physiological and biochemical studies. Degenerative reactions associated with anoxia (Hetherington *et al.*, 1987, 1988; Monk *et al.*, 1989), desiccation (Senaratna *et al.*, 1987) and freezing (Kendall and McKersie, 1989) are similar to those caused by the herbicide paraquat (Bowler *et al.*, 1991; Chia et al., 1982), and the pollutant ozone (Van Camp *et al.*, 1994). Secondly, microsomal membranes from acclimated plants are more tolerant of *in vitro* free radical treatment than those from non-acclimated plants (Kendall and McKersie, 1989). Thirdly, as plants acclimate at low temperatures, they acquire coincidentally increased tolerance to freezing stress, ice-encasement stress and free radical-generating herbicides (Bridger *et al.*, 1994).

We hypothesized that by enhancing a plant's tolerance of oxidative stress we would improve its ability to survive the combination of stresses associated with winter. The mechanisms to detoxify oxygen radicals are varied and the complex interactions among the antioxidants in different subcellular compartments, cells and tissues are only now being elucidated (e.g. Allen, 1995; Bowler *et al.*, 1992; Foyer *et al.*, 1994; Herouart, 1993; Scandalios, 1993). Superoxide dismutase (SOD) is an essential component of these defence mechanisms because it dismutates two superoxide radicals to produce hydrogen peroxide and oxygen. Previously, a Mn-SOD cDNA from *Nicotiana plumbaginifolia* was introduced into alfalfa (*Medicago sativa* L.); one of the primary transformants and its F1 transgenic progeny showed increased survival and vigour after exposure to sub-lethal freezing stress in the laboratory (McKersie *et al.*, 1993). Two of the transgenic plants also had relatively increased tolerance of water deprivation and four had increased vigour and winter survival in the field (McKersie *et al.*, 1996). Our

---

Plant Responses to Environmental Stress, edited by M.F. Smallwood, C.M. Calvert and D.J. Bowles.
© 1999 BIOS Scientific Publishers, Oxford.

previous results suggested that we might modify a physiological process, even as complex as winter-hardiness, by modifying the expression of a single gene using a genetic engineering approach. This encouraged us to conduct a more thorough study.

2. Characterization of transgenic alfalfa plants expressing superoxide dismutase

In this study we have expanded our original observations by transforming two elite alfalfa plants, designated N4 and S4. Petiole explants of *Medicago sativa* were co-cultivated with an overnight culture of *Agrobacterium tumefaciens* C58C1 Rif pMP90 containing the binary vectors pMitSOD or pChlSOD (Bowler *et al.*, 1991). The putatively transgenic plants were screened for the presence of the *nos-nptII* transgene using the polymerase chain reaction (PCR) and by Southern hybridization. SOD activity was determined by staining native polyacrylamide gels with nitroblue tetrazolium and riboflavin (McKersie *et al.*, 1993). The non-transgenic control N4 plant had three major SOD bands – a fast-moving chloroplastic form of Cu/Zn-SOD, a slower moving cytosolic form of Cu/Zn-SOD and a mitochondrial Mn-SOD (*Figure 1*). The transgenic plants had an additional Mn-SOD enzyme superimposed on the native Mn-SOD isozyme in the native gels. In the case of the pMitSOD, the two Mn-SOD forms were not resolved by polyacrylamide gel electrophoresis (PAGE), but in the case of the pChlSOD, two distinct Mn-SOD bands were apparent. The difference in the mobility of the mitochondrial and chloroplast targeted forms of Mn-SOD possibly reflects differences in the cleavage site of the transit peptide, or another post-transcriptional modification.

Mn-SOD activity was increased by a variable amount among the independent transgenic plants containing either pMitSOD or pChlSOD. In about 25% of the transgenic plants from either transformation vector, Mn-SOD activity was reduced or not changed. In the majority of the transgenic plants, Mn-SOD activity was increased less than two-fold. In only a small proportion of the transgenic plants, Mn-SOD activity was doubled relative to total SOD activity.

The highest expressing plants were selected for more detailed analysis. In three pMitSOD and two pChlSOD plants, Mn-SOD activity was increased in all shoot tissues sampled but the increase was not uniform across tissues. The increase in Mn-SOD activity was least in apex, nodes and stem and greatest in the leaf tissues, both petiole and blade. The response was similar in both vectors. The variation among tissues in the transgenic plants was possibly due to differences in transcription of the CaMV35S promoter among the tissues, but post-transcriptional modifications of SOD activity are also likely to be important contributing factors.

The plants containing pMitSOD had the same activity of both forms of Cu/Zn-SOD as the N4 control; therefore total SOD activity in leaf extracts increased (*Table 1*). In contrast, targeting the Mn-SOD to the chloroplast caused an apparent feedback regulation leading to lower chloroplastic and cytosolic Cu/Zn-SOD activity; therefore, total SOD activity was actually reduced even though Mn-SOD was increased. Although the

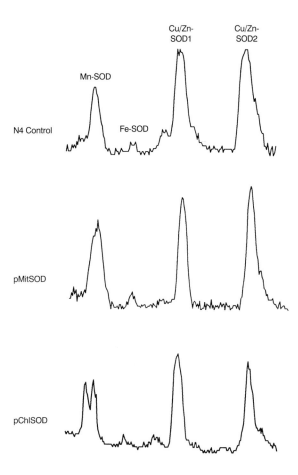

Figure 1. Line-scan of native PAGE gels for superoxide dismutase (SOD) activity in leaf extracts of independent transgenic *Medicago sativa* plants expressing pMitSOD or pChlSOD. From McKersie, B.D., Bowley, S.R. and Jones, K.S. (1999) Winter survival of transgenic Alfalfa overexpressing superoxide dismutase. *Plant Physiology*, vol. 119, pp. 839–847. Reprinted by permission of The American Society of Plant Physiologists.

mechanism of this apparent regulation is unknown, there appeared to be similar regulation of the two Cu/Zn forms of the enzyme.

Four highly expressing primary transgenic plants from each vector were selected for growth analysis and stress tolerance testing in the greenhouse. In the absence of a freezing stress, three of the four pMitSOD plants tested had greater shoot dry matter production than the control. Two pChlSOD plants had less growth and two had more growth than the control.

Table 1. Mn- and Cu/Zn-superoxide dismutase activities (units g^{-1} protein) in independent transgenic *Medicago sativa* plants expressing pMitSOD or pChlSOD

	Mn-SOD	Cu/Zn-SOD1	Cu/Zn-SOD2	Total
N4	426	883	968	2278
N4 pMitSOD	1111*	759 NS	996 NS	2851*
N4 pChlSOD	802*	371*	667*	1840*

Activities were measured on three high-expressing pMitSOD plants, two high-expressing pChlSOD plants and the non-transgenic N4. Values averaged across eight tissues, including shoot apex, leaves 1–4, petioles, nodes and internodes; $n = 50$ (control), 106 (pMitSOD), 58 (pChlSOD).
*Indicates a significant difference between transgenic and control plants at $p = 0.05$ according to a protected LSD test; NS indicates not significantly different.
Data modified from McKersie, B.D., Bowley, S.R. and Jones, K.S. (1999) Winter survival of transgenic Alfalfa overexpressing superoxide dismutase. *Plant Physiology*, vol. 119, pp. 839–847. Reprinted by permission of The American Society of Plant Physiologists.

3. Effect of superoxide dismutase over-expression on freezing tolerance and winter survival

To measure freezing tolerance in the greenhouse, the plants were acclimated at 2°C to three different stages. Leaf blades and fibrous roots were frozen to four different temperatures. Statistically significant effects were observed for tissues, acclimation and temperatures (*Table 2*). Roots were consistently more freezing sensitive than leaves. Both leaves and roots had less injury with acclimation. Lower temperatures caused increasing amounts of injury. Small statistically significant differences were observed among the transgenic plants and in comparison with the control plant. However, some transgenic plants had less injury than the non-transgenic N4 control, whereas others had more injury. Nonetheless, the most tolerant transgenic plant was only 1°C more freezing tolerant than the control, indicating that any improvement in freezing tolerance *per se* was minimal.

A field trial was established in spring 1996 by transplanting rooted cuttings of each transgenic and control genotype. The test was arranged in a randomized complete block design with 15 cuttings of each control (non-transgenic) and five cuttings of each transgenic genotype as the experimental units and three replications (blocks). Plants were harvested once in the year of transplanting. Stand counts were taken in autumn 1996 and spring 1997 to determine survival. All plants had good survival in the year of transplanting and entered the winter of 1996/97 with nearly 100% survival. The location of the field trial experienced a relatively harsh winter in 1996/97 because of excessively wet soil conditions. Survival in spring 1997 was less than commonly observed with these parental clones in other tests. The average survival of the transgenic plants was higher than the non-transgenic controls (*Table 3*). The N4 group of transgenic plants had higher survival and vigour in the field than the S4 group, but the two vectors, pMitSOD and pChlSOD were similar in both.

Plants were defoliated on 28 June, 28 July and 28 August 1997 to determine dry matter yields. The average yield of the transgenic plants expressing pMitSOD was

Table 2. Electrolyte leakage (% of total) from leaves and roots after freezing of independent transgenic *Medicago sativa* plants expressing pMitSOD or pChlSOD

Vector	Plant	Leaf	Root
Control	N4	51	68
pMitSOD	N4 pMitSOD-1	53	67
	N4 pMitSOD-6	52	66
	N4 pMitSOD-9	48	66
	N4 pMitSOD-10	47	66
pChlSOD	N4 pChlSOD-1	47	64
	N4 pChlSOD-6	44	66
	N4 pChlSOD-7	46	70
	N4 pChlSOD-9	45	68
LSD (0.05)		0.6 ($n = 24$)	0.5 ($n = 36$)

The experiment was a factorial experiment with a split plot design of three acclimation stages; plants were acclimated at 2°C for 2 weeks or 4 weeks, or acclimated for 4 weeks, frozen at −2°C, then acclimated at 2°C for one more week. All plants were frozen to four temperatures of −6, −7, −8 or −9°C, thawed at 2°C overnight and conductivity of a 10 ml leachate measured. Leaf and root samples were analysed separately. The values shown are the genotype main effects averaged across temperature and acclimation treatments. LSD (0.05) is the least significant difference at the 5% level of probability.

Data modified from McKersie, B.D., Bowley, S.R. and Jones, K.S. (1999) Winter survival of transgenic Alfalfa overexpressing superoxide dismutase. *Plant Physiology*, vol. 119, pp. 839–847. Reprinted by permission of The American Society of Plant Physiologists.

higher than the N4 and S4 non-transgenic controls (*Table 4*). The average yield of those expressing pChlSOD was the same in the N4 group and significantly greater in the S4 group; although in the latter case there was only one transgenic plant tested.

For both survival and yield, there was more variation observed within each group of transgenic plants than there was between parent clones or vectors. The N4 non-transgenic control had 53% survival, and S4 had 29%. Some individual transgenic plants had 100% survival, but there were other individual plants with less survival than the non-transgenic controls in both parental groups and with both vectors (see *Table 3*).

Table 3. Percentage survival of independent transgenic *Medicago sativa* plants expressing pMitSOD or pChlSOD in spaced plantings at Elora, Ontario

Clone	Vector	Number of independent transgenics	Average survival (%)	Maximum survival (%)	Minimum survival (%)
N4	Control	1	53		
	pMitSOD	10	69	40	93
	pChlSOD	9	69	47	100
S4	Control	1	29		
	pMitSOD	13	50	0	87
	pChlSOD	1	67		

Two parent clones designated as N4 and S4 were transformed with the pMitSOD or pChlSOD vectors. Average survival (%) is the average across all independent transgenic plants within a vector with three replications per independent transgenic plant; maximum and minimum are the independent transgenic plants with the highest and lowest survivals.

Data modified from McKersie, B.D., Bowley, S.R. and Jones, K.S. (1999) Winter survival of transgenic Alfalfa overexpressing superoxide dismutase. *Plant Physiology*, vol. 119, pp. 839–847. Reprinted by permission of The American Society of Plant Physiologists.

Table 4. Yield (g/plant) of independent transgenic *Medicago sativa* plants expressing pMitSOD or pChlSOD in spaced plantings at Elora, Ontario

Clone	Vector	Number of independent transgenics	Average yield (g/plant)	Maximum yield (g/plant)	Minimum yield (g/plant)
N4	Control	1	34		
	pMitSOD	10	50	20.2	70.5
	pChlSOD	9	36	17.8	79.3
S4	Control	1	15		
	pMitSOD	13	19	0	33.8
	pChlSOD	1	52		

Yield is total herbage yield in the production year after the first winter of the plants described in Table 3. Average yield (g/plant) is the average across all independent transgenic plants within a vector with three replications per independent transgenic plant; maximum and minimum are the independent transgenic plants with the highest and lowest yields.

Data modified from McKersie, B.D., Bowley, S.R. and Jones, K.S. (1999) Winter survival of transgenic Alfalfa overexpressing superoxide dismutase. *Plant Physiology*, vol. 119, pp. 839–847. Reprinted by permission of The American Society of Plant Physiologists.

Similarly, although both pMitSOD and pChlSOD had individuals with dramatic improvements in yield, approaching twice the yield of the N4 non-transgenic control, other individuals had lower yields than the non-transgenic controls (see *Table 4*). There was only a slight correlation between the yield of transgenic plants in the greenhouse and in the field (*Figure 2*) suggesting that pre-screening in the greenhouse for performance is not effective.

The following year, a second field trial was established by transplanting rooted cuttings of each transgenic and control genotype in 1 × 2 m rectangular plots at 100 plants per plot. Each plot consisted of a population of independent transgenic plants for each construct. Crowns and roots of these field-acclimated alfalfa plants were subjected to freezing temperatures by placing the plants in moist filter paper. The samples were frozen to −6, −8, −10, −12 or −14°C and assessed by viability staining with tetrazolium (Tanino and McKersie, 1985). No differences in the sites or pattern of freezing injury were found between the control and any of the transgenic plants (data not shown). These observations confirm that the transgenic plants are not more freezing tolerant *per se* than the control plant after acclimation in the field.

4. Effect of superoxide dismutase over-expression on bacterial and fungal wilt resistance

Another possible explanation for the improvement in winter survival was that the transgenic plants were more tolerant of fungal or bacterial diseases. Two of the major diseases of alfalfa in North America are *Verticillium* wilt and bacterial wilt. To test for tolerance of these diseases, cuttings of the primary transgenic plant were rooted and infected with each pathogen using standard procedures. Tolerance was based on a visual

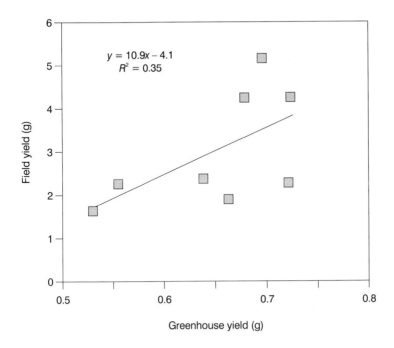

Figure 2. Correlation between field and greenhouse yields of independent transgenic *Medicago sativa* plants expressing pMitSOD or pChlSOD.

assessment of injury. Surprisingly, instead of being more tolerant of the diseases, all the transgenic plants tested in our preliminary experiment were more susceptible than the N4 control plant (*Figure 3*). However, more detailed experiments are required to confirm that increasing SOD activity in the transgenic plant has interfered with its natural defence response and allowed more extensive development of injury symptoms.

5. Discussion

We can draw several conclusions from these data. Firstly, we must be cautious in interpreting experiments that measure stress tolerance effects in transgenic plants that use only a few independent transgenic plants from each vector, such as we have done previously (McKersie *et al.*, 1996), because conflicting results are likely. For example, in the case of SOD over-expression, Pitcher *et al.* (1991), Tepperman and Dunsmuir (1990) and Payton *et al.* (1997) found no improvement in tolerance to oxidative and related stresses, while Gupta *et al.* (1993a, 1993b), Bowler *et al.* (1991), Van Camp *et al.* (1994, 1996), Perl *et al.* (1993) and McKersie *et al.* (1993, 1996) found significant improvements. In several studies (e.g. Gupta *et al.*, 1993a, 1993b; Pitcher *et al.*, 1991) only two transgenic plants were evaluated. The present data would suggest that the required number of independent transgenic alfalfa plants should be over 20 before a reliable prediction of a general trend can be made.

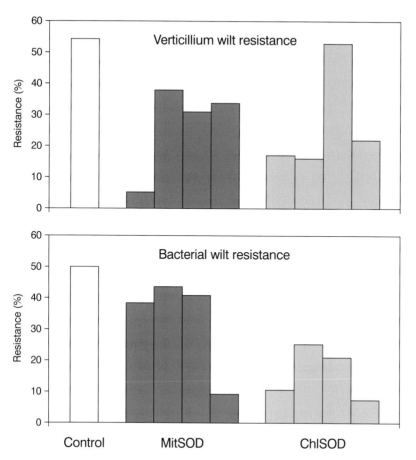

Figure 3. Verticillium and bacterial wilt resistances of independent transgenic *Medicago sativa* plants expressing pMitSOD or pChlSOD.

Secondly, the transgenic alfalfa plants had considerable variation in SOD activity, so we attempted to select plants in the greenhouse to reduce the number of primary transgenic plants that would be evaluated in the field and to speed the selection of plants to be used in cross-pollination for seed production. Pre-screening the transgenic plants for SOD activity, vigour or freezing tolerance in the greenhouse was not effective at identifying individual transgenic plants with improved field performance.

Thirdly, freezing tolerance was increased by only 1°C in the best transgenic plant when the plants were acclimated in controlled environments and when viability was measured by electrolyte leakage. When the crowns and taproots of field-acclimated plants were frozen, there was no apparent difference in freezing tolerance between transgenic and control plants when viability was measured by tetrazolium staining. Therefore, there is no indication from these studies that over-expression of SOD changed the primary site of freezing injury in alfalfa. Yet differences in winter survival were observed among the plants.

Finally, SOD expression has complex effects on the physiology of the plant; apparently SOD over-expression in these plants increased their winter survival but decreased tolerance of *Verticillium* and bacterial wilts. This may explain why plant breeding programmes that have targeted both improved persistence and improved disease tolerance as objectives have failed to select for naturally occurring plants with increased SOD activity.

The mechanism by which SOD over-expression increased winter survival is still not clear. Additional SOD activity may have improved the cellular repair mechanisms allowing the transgenic plant to better recover from freezing injury. In addition, freezing tolerance may not have limited winter survival and better results may have been obtained with ice-encasement, flooding or long duration freezing tests. Unfortunately at the present time, despite the regulatory requirements for federal permits to test transgenic plants in the field, field-testing remains the most suitable measure for winter survival and yield potential, even for transgenic plants.

Acknowledgements

The binary vectors pMitSOD and pChlSOD were kindly provided by Dirk Inzé, Universiteit Gent, Belgium. The transformation of the N4 alfalfa plants was done by Molian Deng, and the S4 plants by Ranjith Pathirana. Lori Wright confirmed the transformations using PCR and conducted the Southern hybridizations. Heather Anderson, Andrea Fiebig and Carol Hannam analysed SOD activity using PAGE. Ning Chen maintained the plants in the greenhouse and conducted the freezing experiments. Donna Hancock, Jennifer Thatcher and Julia Murnaghan conducted the field trials.

References

Allen, R.D. (1995) Dissection of oxidative stress tolerance using transgenic plants. *Plant Physiol.* **107**: 1049–1054.

Bowler, C., Slooten, L., Vandenbranden, S., de Rycke, R., Botterman, J., Sybesma, C., van Montagu, M. and Inzé, D. (1991) Manganese superoxide dismutase can reduce cellular damage mediated by oxygen radicals in transgenic plants. *EMBO J.* **10**: 1723–1732.

Bowler, C., Van Montagu, M. and Inzé, D. (1992) Superoxide dismutase and stress tolerance. *Annu. Rev. Plant Physiol. Plant Mol. Biol.* **43**: 83–116.

Bridger, G.M., Yang, W., Falk, D.E. and McKersie, B.D. (1994) Cold acclimation increases tolerance of activated oxygen in winter cereals. *J. Plant Physiol.* **144**: 235–240.

Chia, L.S., McRae, D.G. and Thompson, J.E. (1982) Light-dependence of paraquat-initiated membrane deterioration in bean plants. Evidence for the involvement of superoxide. *Physiol. Plant* **56**: 492–499.

Foyer, C.H., Descourvieres, P. and Kunert, K.J. (1994) Protection against oxygen radicals: an important defence mechanism studied in transgenic plants. *Plant Cell Environ.* **17**: 507–523.

Gupta, A.S., Heinen, J.L., Holaday, A.S., Burke, J.J. and Allen, R.D. (1993a) Increased resistance to oxidative stress in transgenic plants that overexpress chloroplastic Cu/Zn superoxide dismutase. *Proc. Natl Acad. Sci. USA* **90**: 1629–1633.

Gupta, A.S., Webb, R.P., Holaday, A.S. and Allen, D. (1993b) Over-expression of superoxide dismutase protects plants from oxidative stress. Induction of ascorbate peroxidase in superoxide dismutase-overexpressing plants. *Plant Physiol.* **103**: 1067–1073.

Herouart, D., Bowler, C., Willekens, H., Van Camp, W., Slooten, L., Van Montagu, M. and Inzé, D. (1993) Genetic engineering of oxidative stress resistance in higher plants. *Phil. Trans. Roy. Soc. Lond. Series B, Biol. Sci.* **342**: 235–240.

Hetherington, P.R., McKersie, B.D. and Borochov, A. (1987) Ice-encasement injury to microsomal membranes from winter wheat crowns I. Comparison of membrane properties after lethal ice-encasement and during a post-thaw period. *Plant Physiol.* **85**: 1068–1072.

Hetherington, P.R., Broughton, H.L. and McKersie, B.D. (1988) Ice-encasement injury to microsomal membranes from winter wheat crowns II. Changes in membrane lipids during ice-encasement. *Plant Physiol.* **86**: 740–743.

Kendall, E.J. and McKersie, B.D. (1989) Free radical and freezing injury to cell membranes of winter wheat. *Physiol Plant.* **76**: 86–94.

McKersie, B.D., Chen, Y., de Beus, M., Bowley, S.R., Bowler, C., Inzé, D., D'Halluin, K. and Botterman, J. (1993) Superoxide dismutase enhances tolerance of freezing stress in transgenic alfalfa (*Medicago sativa* L.) *Plant Physiol.* **103**: 1155–1163.

McKersie, B.D., Bowley, S.R., Harjanto, E. and Leprince, O. (1996) Water deficit tolerance and field performance of transgenic alfalfa overexpressing superoxide dismutase. *Plant Physiol.* **111**: 1177–1181.

Monk, L.S., Fagerstedt, K.V. and Crawford, R.M.M. (1989) Oxygen toxicity and superoxide dismutase as an antioxidant in physiological stress. *Physiol Plant.* **76**: 456–459.

Payton, P., Allen, R.D., Trolinder, N. and Holaday, A.S. (1997) Over-expression of chloroplast-targeted Mn superoxide dismutase in cotton (*Gossypium hirsutum* L., cv. Coker 312) does not alter the reduction of photosynthesis after short exposures to low temperature and high light intensity. *Photosynth Res.* **52**: 233–244.

Perl, A., Perl-Treves, R., Galili, S., Aviv, D., Shalgi, E., Malkin, S. and Galun, E. (1993) Enhanced oxidative-stress defense in transgenic potato expressing tomato Cu,Zn superoxide dismutases. *Theor. Appl. Genet.* **85**: 568–576.

Pitcher, L.H., Brennan, E., Hurley, A., Dunsmuir, P., Tepperman, J.M. and Zilinskas, B.A. (1991) Overproduction of petunia chloroplastic copper/zinc superoxide dismutase does not confer ozone tolerance in transgenic tobacco. *Plant Physiol.* **97**: 452–455.

Scandalios, J.G. (1993) Oxygen stress and superoxide dismutases. *Plant Physiol.* **101**: 7–12.

Senaratna, T., McKersie, B.D. and Borochov, A. (1987) Desiccation and free radical mediated changes in plant membranes. *J Exp. Bot.* **38**: 2005–2014.

Tanino, K.K. and McKersie, B.D. (1985) Injury within the crown of winter wheat seedlings after freezing and icing stress. *Can. J. Bot.* **63**: 432–436.

Tepperman, J.M. and Dunsmuir, P. (1990) Transformed plants with elevated level of chloroplastic SOD are not more resistant to superoxide toxicity. *Plant Mol. Biol.* **14**: 501–511.

Van Camp, W., Willekens, H., Bowler, C., van Montagu, M. and Inzé, D. (1994) Elevated levels of superoxide dismutase protect transgenic plants against ozone damage. *Bio/Tech* **12**: 165–168.

Van Camp, W., Capiau, K., Van Montagu, M., Inzé, D. and Slooten, L. (1996) Enhancement of oxidative stress tolerance in transgenic tobacco plants overproducing Fe-superoxide dismutase in chloroplasts. *Plant Physiol.* **112**: 1703–1714.

The technological potential of biological frost protection – or if only we were as clever as plants

Peter J. Lillford and Chris B. Holt

1. Background

Recent years have seen a dramatic growth in the use of freezing as a preservation process for foods. Freezing has the advantage of limiting microbial and chemical spoilage, whilst not causing the deleterious effects on flavour and texture associated with other preservation methods such as drying or sterilization. Unfortunately, the freezing process still causes structural damage in tissues due to the freeze-concentration of solutes, the slow growth of ice crystals at most commercial storage temperatures, and the rapid structural damage which occurs on thawing. The problems of food freezing are quite comparable to those involved in prolonged organ preservation and even less routes to its preservation are allowed. Nature, however, appears to have the solution since many plants are able not only to survive freezing, but subsequently return to normal growth and viability. They appear to have solved the problem of protecting subcellular structures and metabolism. Not surprisingly, we would like the secret. The identification of specific antifreezes and ice nucleators, as well as the increase of normal freezing point-depressing agents such as sucrose, has stimulated interest in their source, structure and technical use. Much of the information appears in the patent literature before detailed scientific papers are published. We will review the state of the art.

Raoult's Law teaches us that solutes depress the freezing point of water and that low molecular mass compounds are most effective. The addition of small solutes is common practice to reduce ice content in frozen foods, but frost-tolerant plants also know the secret, and accumulate small sugars and amino acids in their cytosol. It is also well known to physicists that water in small droplets are less likely to contain heterogeneous nucleation sites so that stable supercooling in small plant cells is also possible.

What we do not properly understand is whether the small molecules which plants accumulate have any other specific stabilizing effects on membranes or dissolved protein, though sugars are generally capable of stabilizing proteins as their concentration increases.

But there are a lot more tricks that Nature has evolved to protect living organisms

---

Plant Responses to Environmental Stress, edited by M.F. Smallwood, C.M. Calvert and D.J. Bowles.
© 1999 BIOS Scientific Publishers, Oxford.

from cold and ice. We review here the special cryoprotective molecules, and the first hesitant steps of technologists to develop their potential.

2. The glassy state

If ice forms outside the cell, internal solute concentrations rise and the viscosity of the cytosol increases. We now know that at sufficiently high concentrations, the solution can form an amorphous glass, in which diffusion of the solutes stops and therefore so do chemical reactions, so that the crystallization of water to ice takes many days or months (Levine and Slade, 1988). For small sugars this is at a concentration of 80 wt%, but the temperature at which this is reached is dependent on the solute type. In general, higher molecular mass solutes produce a glass closer to 0°C, so there is a compromise between freezing point depression and glassing. Our knowledge of the glassing behaviour of complex mixtures of solutes like those of the cytosol is still very limited.

3. Nucleators: food applications

Freeze-concentration is a much milder process than removing water by evaporation. Watanabe and co-workers have published a series of papers describing the use of bacterial ice nucleators in freeze-concentration (Honma et al., 1993; Kumeno et al., 1993, 1994; Watanabe and Arai, 1987; Watanabe et al., 1989, 1991). Ice nucleators reduce the number of ice crystals and increase their size which makes it easier to separate the ice from the unfrozen liquid phase. These workers freeze-concentrated egg white in a beaker which had ice-nucleating bacteria attached to the wall. Foams which were subsequently made with the concentrated egg white were superior to those made with intact egg white. Lemon juice was also freeze-concentrated (Watanabe et al., 1989) and the volatiles were well retained compared to vacuum concentration.

Watanabe and Arai (1987) experimented with the use of ice-nucleating bacteria in freeze-drying of food. When ice-nucleating bacteria were added to soy sauce and soybean oil the freeze-dried products had a porous structure with large cavities which were easier to reduce to a powder. Nagashima and Suzuki (1985) found that high-salt foods could only be frozen after dilution with water because of the very low nucleation temperature. However, addition of ice-nucleating bacteria initiated freezing without the need to dilute.

Freeze-texturing is sometimes used to produce meat-like textures from plant material. However, freezing is usually preceded by supercooling and when freezing then takes place it gives a disordered ice formation which produces an isotropic texture. Arai and Watanabe (1986) found that adding bacterial ice nucleators minimized the supercooling and gave long parallel ice crystals. Directional textures were produced in raw egg white, soybean protein isolate and cornstarch paste. Jann et al. (1997) used nucleators from sea buckthorn berries to freeze-texture a soya bean protein isolate. They also used the nucleators from sea buckthorn berries to elevate the freezing temperature of ice-cream, meat paste and coffee extracts.

Li et al. (1997) added ice-nucleating bacteria to model foods and froze them in air at a temperature of −6° to −7°C. The addition of bacteria raised the nucleation temperatures of the model foods and so reduced the total freezing time by between 20 and 38%. The authors suggest this would lead to energy savings and improvements in efficiency and product quality.

Ice-nucleating bacteria have been used as a rapid test for *Salmonella* in foods (Worthy, 1990). A *Salmonella*-specific bacteriophage is incorporated with a gene coding for the ice-nucleation protein. The bacteriophage reagent and a green fluorescent dye are then added to samples of the food so that any *Salmonella* present will be infected with the bacteriophage. The samples are then incubated at 37°C and then cooled to −5°C. If bacteria are present then they will now nucleate ice and the samples will freeze and the dye will change colour.

4. Nucleators: non-food applications

The use of ice-nucleating bacteria in artificial snow-making is described by Woerpel (1980). The bacteria are added to water which is then sprayed into the atmosphere as fine droplets when the air temperature is below freezing. The bacteria increase the nucleation temperature of the droplets so that artificial snow can be made at higher ambient temperatures and higher humidities and the likelihood of unfrozen droplets falling to the ground and producing poor quality snow is reduced.

Lee (1997) describes a novel way of destroying insect pests, especially those found in stored grain. Most of the pests are freeze-avoiding and will have a low freezing temperature in the winter. If these pests are treated with ice-nucleating bacteria such as *Pseudomonas syringae* then when the ambient temperature falls in the winter the bacteria nucleate ice and the pests freeze and die.

There are many patents which relate ice-nucleating bacteria to frost protection of high value crops and ornamental plants. Lindow (1983) claims that treating plants with antagonistic bacteria which are capable of diminishing the availability of a limited nutrient provided by the host plant will inhibit colonization by ice-nucleating bacteria. Arny and Lindow (1979) describe treating frost-sensitive plants with non-ice-nucleating bacteria to reduce the population of ice-nucleating bacteria normally present. Lindow *et al.* (1985) describe bacteria with diminished ice-nucleating activity as a result of genetic modification which can be used for that purpose.

Spray-ice technology is used in Arctic regions for the construction of load-bearing structures such as aircraft runways, roads and ice islands for offshore drilling. The construction of ice structures can only be carried out when the air temperatures are sufficiently low to cause rapid freezing. However, the ice-nucleating bacteria *Pseudomonas syringae* has successfully been applied in Arctic ice construction to effectively reduce the supercooling of sprayed seawater (Owen *et al.*, 1987).

5. Antifreezes: food applications

In the recrystallization process the large ice crystals in a population grow at the expense of the small crystals. Extremely low concentrations of antifreeze proteins (AFPs) have been found to inhibit recrystallization of ice. For example, Knight *et al.* (1984) showed that AFGP (antifreeze glycoprotein) can inhibit ice recrystallization at a concentration of 100 µg l^{-1}, a level that is too low to cause any thermal hysteresis. Warren *et al.* (1990) produced antifreeze polypeptides and their fusion proteins by chemical and recombinant DNA techniques which were able to prevent or slow down ice recrystallization. Lillford *et al.* (1998) described antifreeze proteins extracted from grass which are capable of preventing recrystallization even after heating for 10 min at 100°C.

AFPs are also able to minimize the size of ice crystals which form when freezing first takes place. When large ice crystals form in meat or fish tissues they can damage the membranes and cause drip loss during thawing. For this reason Payne *et al.* (1994) soaked meat (bovine and ovine muscle) in solutions of AFP I or AFGP and then froze at −20°C. The samples showed evidence of reduced ice crystal size and reduced tissue damage compared with the controls.

Darling *et al.* (1998) isolated antifreeze peptides from many plant species and showed that, when added to ice-cream, they were capable of keeping the ice crystals to a size below 15 µm after quick freezing to −40°C. They claimed this would improve the freezing tolerance of frozen products and improve the texture of ice-cream.

A very simple procedure for making ice-cream whilst still retaining small ice crystals is described by Clemmings *et al.* (1996). The ice-cream is prepared containing AFPs and is prepared merely by cooling without any stirring. The antifreeze keeps the crystals at a small size in this process.

Because antifreeze proteins have preferred binding planes on the ice crystal lattice they can be used to control the shape of ice crystals. Needham *et al.* (1998a) used AFPs which bind preferentially to the primary prism planes on the ice crystal lattice so as to produce elongated ice crystals. The process is used in the production of frozen confectionery and ice-cream. The ice crystals are not prone to recrystallization and give ice cream with a relatively hard and brittle texture which is desirable for certain applications.

If the antifreeze proteins bind preferentially to the pyramidal planes in the ice crystal lattice then the ice crystals have a compact rounded shape. Needham *et al.* (1998b) describe the use of such ice crystals to give soft ice-cream and also describe combining this ice-cream with brittle ice-cream containing elongated crystals to give a product with texture contrast.

AFPs can apparently reduce the rate of growth of ice in plant tissue during freezing. Cutler *et al.* (1989) added winter flounder AFP to bromegrass *Bromis inermis* cell suspensions and then froze them. They found that the rate of freezing was lower than that in an untreated cell suspension and the amount of ice was also lower at any given temperature. Hightower *et al.* (1991) produced transgenic tomato plants that accumulate a protein A-type AFP fusion protein in the leaves in sufficient quantities to reduce the rate of ice recrystallization.

AFPs have been used to avoid freezing in fish tissues. The lethal freezing temperature of rainbow trout was lowered by direct injection of type I AFP and the reduction was proportional to the amount which was injected (Cutler *et al.*, 1989). Hew *et al.* (1995) describe transgenic Atlantic salmon *Salmo solar* which produce AFP I from introduced flounder genes. The intention is to protect the salmon from freezing in shallow coastal waters but so far the expression levels are too low to achieve this.

6. Antifreezes: non-food applications

In some cases it has been found that AFPs can be useful in cryopreservation. Rubinsky *et al.* (1992) studied the freezing of immature pig oocytes and found that addition of AFGPs from Antarctic notothniid fishes dramatically improved both the survival and the morphological integrity during rapid cooling to cryogenic temperatures.

Liposomes normally leak about 50% of their contents when they are cooled through the transition temperature. However, addition of less than 1% AFGP can prevent up to 100% of this leakage both during chilling and warming through the phase transition (Hays *et al.*, 1996).

Under certain circumstances AFPs can increase the damage which occurs during freezing which may be useful in surgical applications. Koushafar *et al.* (1997) perfused rat livers with saline solutions with and without 10 mg ml^{-1} of AFP I and then froze them. The AFP significantly increased the cellular destruction apparently through formation of intracellular ice. It was concluded that antifreeze proteins may be effective chemical adjuvants to cryosurgery.

Water-ice mixtures are commonly used as thermal storage systems for applications such as air conditioning. Such systems produce savings in energy and maintenance costs and can also have lower capital costs. Grandum and Nakagomi (1997) describe the addition of AFPs to ice storage systems to improve the flowability by preventing recrystallization and controlling ice crystal shape.

References

Arai, S. and Watanabe, M. (1986) Freeze-texturing of food materials by ice nucleation with the bacterium *Erwinia ananas*. *Agric. Biol. Chem. Tokyo* **50**: 169–175.

Arny, D.C. and Lindow, S.E. (1979) Method for reducing the temperature at which plants freeze. Patent US 4161084.

Clemmings, J.F., Huang, V.T., Rosenwald, D.R. and Zoerb, H.F. (1996) Frozen compositions with antifreeze protein additives. Patent WO 9639878.

Cutler, A.J., Saleem, M., Kendall, E., Gusta, L.V., Georges, F. and Fletcher, G.L. (1989) Winter flounder antifreeze protein improves the cold hardiness of plant tissues. *J. Plant Physiol.* **135**: 351–354

Darling, D.F., Doucet, C.J., Fenn, R.A., Lillford, P.J., McArthur, A.J., Needham, D., Sidebottom, C.M., Worrell, D. and Byass, L.J. (1998) Frozen confectionery products containing at least one antifreeze protein derived from plants, Patent WO 9804148.

Grandum, S. and Nakagomi, K. (1997) Characteristics of ice slurry containing antifreeze protein for ice storage applications. *J. Thermophysics Heat Transfer* **11**: 461–466.

Hays, L.M., Feeney, R.E., Crowe, L.M., Crowe, J.H. and Oliver, A.E. (1996) Antifreeze glycoproteins inhibit leakage from liposomes during thermotropic phase transitions. *Proc. Natl Acad. Sci. USA* **93**: 6835–6840.

Hew, C.L., Fletcher, G.L. and Davies, P.L. (1995) Transgenic salmon: tailoring the genome for food production. *J. Fish Biol.* **47**: 1–19.

Hightower, R., Baden, C., Pentes, E., Lund, P. and Dunsmuir, P. (1991) Expression of antifreeze proteins in transgenic plants. *Plant Mol. Biol.* **17**: 1013–1021.

Honma, K., Makino, T., Kumeno, K. and Watanabe, M. (1993) High pressure sterilisation of ice nucleation-active *Xanthamonas campestris* and its application to egg processing. *Biosci. Biotechnol. Biochem.* **57**: 1091–1094.

Jann, A., Lundheim, R., Niederberger, P. and Richard, M. (1997) Obtention of ice nucleating agent from sea buckthorn. Patent US 5637301.

Knight, C.A., DeVries, A.L. and Oolman, L.D. (1984) Fish antifreeze protein and the freezing and recrystallisation of ice. *Nature* **308**: 295–296.

Koushafar, H., Pham, L., Lee, C. and Rubinsky, B. (1997) Chemical adjuvant cryosurgery with antifreeze proteins. *J. Surg. Oncol.* **66**: 114–121.

Kumeno, K., Nakahama, N., Honma, K., Makino, T. and Watanabe, M. (1993) Production and characterisation of a pressure-induced gel from freeze-concentrated milk. *Biosci. Biotechnol. Biochem.* **57**: 750–752.

Kumeno, E., Kurimoto, K., Nakahama, N. and Watanabe, M. (1994) Functional properties of freeze-concentrated egg white foam and its application for food processing. *Biosci. Biotechnol. Biochem.* **58**: 447–450.

Lee, R.E. (1997) Method for destroying insect pests. Patent US 5622698.

Levine, H. and Slade, L. (1988) Thermomechanical properties of small carbohydrate water glasses and rubbers. *J. Chem. Soc. Faraday Trans 1* **84**: 2619–2633.

Li, J., Izquierdo, M. and Lee, T.-C. (1997) Effects of ice-nucleation active bacteria on the freezing of some model food systems. *Int. J. Food Sci. Technol.* **32**: 41–49.

Lillford, P.J., McArthur, A.J., Sidebottom, C.M. and Wilding, P. (1998) Frozen food product containing heat stable antifreeze protein. Patent WO 98/04699.

Lindow, S.E. (1983) Inhibiting plant frost injury using bacteria antagonistic to ice-nucleating bacteria. Patent EP 74718.

Lindow, S.E., Orser, C.S. and Panapoulos, N.J. (1985) Micro-organism with diminished ice-nucleation activity. Patent EP 138426.

Nagashima, N. and Suzuki, E. (1985) Freezing curve by broadline pulsed NMR and freeze-drying. *Refrigeration Sci. Technol.* **1**: 65–70.

Needham, D., Smallwood, K. and Fenn, R.A. (1998a) Frozen food products especially ice cream containing antifreeze peptides. Patent WO 9804146.

Needham, D., Smallwood, K., Darling, D.F. *et al.* (1998b) Antifreeze proteins for use in frozen confectionery food products. Patent WO 9804147.

Owen, L.B., Masterson, D.M. and Green, S.J. (1987) Rapid construction of ice structures with chemically treated sea water. Patent US 4637217.

Payne, S.R., Sandford, D., Harris, A. and Young, O.A. (1994) The effect of antifreeze proteins on chilled and frozen meat. *Meat Sci.* **37**: 429–438.

Rubinsky, B., Arav, A. and Devries, A.L. (1992) The cryoprotective effect of antifreeze glycopeptides from Antarctic fishes. *Cryobiology* **29**: 69–79.

Warren, G.J., Mueller, G.M., McKown, R.L. and Dunsmuir, P. (1990) Antifreeze polypeptides. Patent WO 9013571.

Watanabe, M. and Arai, S. (1987) Freezing of water in the presence of the ice nucleating active bacterium *Erwinia ananas* and its application for efficient freeze drying of foods. *Agric. Biol. Chem.* **51**: 557–563.

Watanabe, M., Watanabe. J., Kumeno, K., Nakahama, N. and Arai, S. (1989) Freeze concentration of some foodstuffs using ice-nucleation active bacterial cells entrapped in calcium alginate gels. *Agric. Biol. Chem.* **53**: 2731–2735.

Watanabe, M., Arai, E., Kumeno, K. and Honma, K. (1991) A new method for producing a non-heated jam sample: the use of freeze concentration and high-pressure sterilisation. *Agric. Biol. Chem.* **55**: 2175–2176.

Woerpel, M.D. (1980) Snow making. Patent US 4200228.

Worthy, W. (1990) Bacterial assay exploits ice nucleation research. *Chem. Eng. News* **68**: January 8: 25–26.

Chapter 17

Molecular responses to drought stress in plants: regulation of gene expression and signal transduction

Kazuo Shinozaki, Kazuko Yamaguchi-Shinozaki, Quang Liu, Mie Kasuga, Kazuya Ichimura, Tsuyoshi Mizoguchi, Takeshi Urao, Shinichi Miyata, Kazuo Nakashima, Zabta K. Shinwari, Hiroshi Abe, Yho Sakuma, Takuya Ito and Motoaki Seki

1. Background

Plants respond and adapt to a variety of environmental stresses in order to survive. Among these environmental stresses drought is one of the most adverse factors of plant growth and crop production. Drought stress induces various biochemical and physiological responses in plants. Moreover, a variety of genes have been described that respond to drought at the transcriptional level (Shinozaki and Yamaguchi-Shinozaki, 1997). Their gene products are thought to function in stress tolerance and response. Recently, stress-inducible genes were used to improve stress tolerance of plants by gene transfer. It is important to analyse functions of stress-inducible genes not only for further understanding of molecular mechanisms of stress tolerance and response of higher plants but also for improvement of stress tolerance of crops by gene manipulation.

The plant growth regulator abscisic acid (ABA) is produced under drought conditions and plays important roles in response and tolerance against dehydration. Most of the genes that have been studied to date are also induced by ABA. It appears that dehydration triggers the production of ABA, which, in turn, induces various genes. Several reports have described genes that are induced by dehydration but are not responsive to exogenous ABA treatments. These findings suggest the existence of ABA-independent as well as ABA-dependent signal-transduction cascades between the initial signal of drought stress and the expression of specific genes (Shinozaki and Yamaguchi-Shinozaki, 1997). To understand the molecular mechanisms of gene expression in response to drought stress, *cis*- and *trans*-acting elements that function in ABA-independent and ABA-responsive gene expression by drought stress have been precisely analysed. A variety of transcription factors are involved in stress-responsive gene expression, which suggests the involvement of complex regulatory systems in molecular responses to drought stress. In this chapter we describe recent progress mainly on

Plant Responses to Environmental Stress, edited by M.F. Smallwood, C.M. Calvert and D.J. Bowles.
© 1999 BIOS Scientific Publishers, Oxford.

cis- and *trans*-acting factors involved in ABA-independent gene expression and signal-transduction in drought stress response, especially the roles of a two-component system and mitogen-activated protein kinase (MAPK) cascade in stress signalling.

2. Functions of drought-inducible genes

A variety of genes are induced by drought stress, and functions of their gene products have been predicted from sequence homology with known proteins. Genes induced during drought stress conditions are thought to function not only in protecting cells' dehydration by the production of important metabolic proteins but also in the regulation of genes for signal transduction in the drought stress response (Shinozaki and Yamaguchi-Shinozaki, 1997). Thus, these gene products are classified into two groups (*Figure 1*). The first group includes proteins that probably function in stress tolerance. The second group contains protein factors involved in further regulation of signal transduction and gene expression that probably function in stress response. Existence of a variety of drought-inducible genes suggests complex responses of plants to drought stress. Their gene products are involved in drought stress tolerance and stress responses.

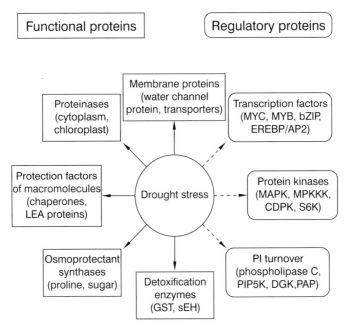

Figure 1. Drought stress-inducible genes and their possible functions in stress tolerance and response. Gene products are classified into two groups. The first group includes proteins that probably function in stress tolerance (Functional proteins), and the second group contains protein factors involved in further regulation of signal transduction and gene expression that probably function in stress response (Regulatory proteins). CDPK, calcium-dependent protein kinase; DGK, diacylglycerol kinase; GST, glutathione-*S*-transferase; LEA, late embryogenesis abundant; MAPK, mitogen-activated protein kinase; PAP, phosphatidic acid phosphatase; PIP5K, phosphatidylinositol-4-phosphate-5-kinase; sEH, soluble expoxide hydrolase; S6K, ribosomal protein S6 kinase.

3. Complex regulatory systems for gene expression by drought stress

The expression patterns of genes induced by drought have been analysed by RNA gel-blot analysis. Results indicated broad variations in the timing of induction of these genes under drought conditions. All the drought-inducible genes are induced by high salinity stress. Most of the drought-inducible genes also respond to cold stress but some of them do not, and *vice versa*. Many genes respond to ABA whereas some others do not. ABA-deficient mutants were used to analyse drought-inducible genes that respond to ABA. Several genes were induced by exogenous ABA treatment, but were also induced by cold or drought in ABA-deficient (*aba*) or ABA-insensitive (*abi*) *Arabidopsis* mutants. These observations indicate that these genes do not require an accumulation of endogenous ABA under cold or drought conditions, but do respond to ABA. There are ABA-independent as well as ABA-dependent regulatory systems of gene expression under drought stress (Shinozaki and Yamaguchi-Shinozaki, 1997). Analysis of the expression of ABA-inducible genes showed that several genes require protein biosynthesis for their induction by ABA, suggesting that at least two independent pathways exist between the production of endogenous ABA and gene expression under stress conditions.

As shown in *Figure 2*, it is now hypothesized that at least four independent signal

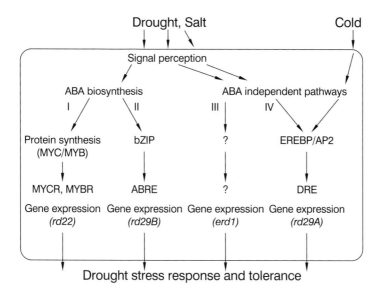

Figure 2. Signal transduction pathways between the perception of drought stress signal and gene expression. At least four signal transduction pathways exist (I–IV): two are abscisic acid (ABA)-dependent (I and II) and two are ABA-independent (III and IV). Protein biosynthesis is required in one of the ABA-dependent pathways (I). In another ABA-dependent pathway, ABA-responsive element (ABRE) functions as an ABA-responsive element and does not require protein biosynthesis (II). In one of the ABA-independent pathways, a drought-responsive element (DRE) is involved in the regulation of genes not only by drought and salt but also by cold stress (IV). Another ABA-independent pathway is controlled by drought and salt, but not by cold (III).

pathways function in the activation of stress-inducible genes under dehydration conditions: two are ABA-dependent (pathways I and II) and two are ABA-independent (pathways III and IV) (Shinozaki and Yamaguchi-Shinozaki, 1998). One of the ABA-dependent pathways requires protein biosynthesis (pathway I). Many stress- and ABA-inducible genes encoding various transcription factors have now been reported. These contain conserved DNA-binding motifs, such as MYB (ATMYB2) and MYC (rd22BP1) (Abe *et al.*, 1997; Urao *et al.*, 1993; Yamaguchi-Shinozaki and Shinozaki, 1993). These transcription factors are thought to function in the regulation of ABA-inducible genes, such as *rd22* in *Arabidopsis*, which respond to drought stress rather slowly after the production of ABA-inducible transcription factors (pathway I).

   Cis- and *trans*-acting factors involved in ABA-induced gene expression have been extensively analysed (pathway II). A conserved sequence, PyACGTGGC, has been reported to function as an ABA-responsive element (ABRE) in many ABA-responsive genes (Shinozaki and Yamaguchi-Shinozaki, 1997). cDNAs encoding DNA-binding proteins that specifically bind to the ABRE have been cloned and shown to contain the bZIP structure. There are several drought-inducible genes that do not respond to either cold or ABA treatment, which suggests that there is a fourth pathway in the dehydration stress response (pathway III). These genes include *rd19* and *rd21* that encode different cysteine proteases (Koizumi *et al.*, 1993), and *ERD1* that encodes a Clp protease regulatory subunit (Kiyosue *et al.*, 1993; Nakashima *et al.*, 1997). Promoter analysis of these genes will give us more information on pathway III. One of the ABA-independent pathways overlaps with that of the cold response (pathway IV). In this chapter, *cis-* and *trans*-acting elements involved in one of the ABA-independent pathway (pathway IV) are precisely described. Recently, based on genetic analysis of *Arabidopsis* mutants with the *rd29A* promoter:luciferase reporter gene, the existence of drought-, salt- and cold-specific signalling pathways in stress response was suggested, but cross-talks between these signalling pathways were also observed (Ishitani *et al.*, 1997).

4.   Important roles of drought-responsive element/C-repeat (DRE/CRT) *cis*-acting element and its DNA-binding proteins in abscisic acid-independent gene expression during drought and cold stress (pathway IV)

A number of genes are induced by drought, salt and cold in *aba* (ABA-deficient) or *abi* (ABA-insensitive) *Arabidopsis* mutants. This suggests that these genes do not require ABA for their expression under cold or drought conditions. Among these genes, the expression of a drought-inducible gene for *rd29A/lti78/cor78* was extensively analysed (Yamaguchi-Shinozaki *et al.*, 1994). At least two separate regulatory systems function in gene expression during drought and cold stress; one is ABA-independent (*Figure 2,* pathway IV) and the other is ABA-dependent (pathway II). A 9-bp conserved sequence, TACCGACAT, named the dehydration-responsive element (DRE), is essential for the regulation of the induction of *rd29A* under drought, low temperature and high salt stress conditions, but does not function as an ABA-responsive element (*Figure 3*). The *rd29A* promoter contains ABRE, which functions in ABA-responsive expression. DRE-related

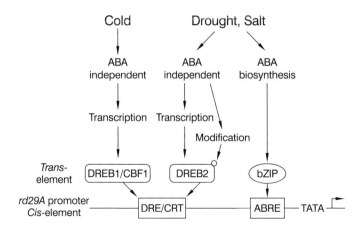

Figure 3. A model of the induction of the *rd29A* gene and *cis-* and *trans*-acting elements involved in stress-responsive gene expression. Two *cis*-acting elements, a drought-responsive element (DRE) and an abscisic acid (ABA) responsive element (ABRE), are involved in the ABA-independent and ABA-responsive induction of *rd29A*, respectively. Two types of different DRE-binding proteins, DREB1 and DREB2, separate two different signal transduction pathways in response to cold and drought stresses, respectively. DREB1s/CBF1 are transcriptionally regulated whereas DRE2s are controlled post-translationally as well as transcriptionally. ABRE-binding proteins encode bZIP transcription factors. CRT, C-repeat.

motifs have been reported in the promoter regions of many cold- and drought-inducible genes. These results suggest that DRE-related motifs including C-repeat (CRT) and low temperature-responsive element (LTRE), which contain a CCGAC core motif, are involved in drought- and cold-responsive but ABA-independent gene expression.

Protein factor(s) that specifically interact with the 9-bp DRE sequence were detected in nuclear extract prepared from either dehydrated or untreated *Arabidopsis* plants (Yamaguchi-Shinozaki and Shinozaki, 1994). Recently, five independent cDNAs for DRE/CRT-binding proteins have been cloned using the yeast one hybrid screening method (Liu *et al.*, 1998). All the DRE/CRT-binding proteins (DREBs and CBFs) contain a conserved DNA-binding motif that has also been reported in EREBP and AP2 proteins (EREBP/AP2 motif) that are involved in ethylene-responsive gene expression and floral morphogenesis, respectively (Liu *et al.*, 1998; Stockinger *et al.*, 1997). These five cDNA clones that encode DRE/CRT-binding proteins are classified into two groups, CBF1/DREB1 and DREB2. Expression of the *DREB1A* gene and its two homologues (*DREB1B=CBF1*, *DREB1C*) was induced by low temperature stress, whereas expression of the *DREB2A* gene and its single homologue (*DREB2B*) was induced by dehydration (Liu *et al.*, 1998; Shinwari *et al.*, 1998). Over-expression of the DREB1A cDNA in transgenic *Arabidopsis* plants not only induced strong expression of the target genes under unstressed conditions but also caused dwarfed phenotypes in the transgenic plants. These DREB1A transgenic plants also revealed freezing and dehydration tolerance, which was also shown in the CBF1 transgenics (Jaglo-Ottosen *et al.*, 1998; Liu *et al.*, 1998). In contrast, over-expression of the DREB2A cDNA induced weak expression of the target genes under unstressed conditions and caused slight growth retardation of the transgenic plants (Liu *et al.*, 1998). These results indicate that two independent families of DREB proteins, DREB1 and DREB2, function as *trans*-acting factors in two separate signal transduction pathways under low temperature and dehydration conditions, respectively (Liu *et al.*, 1998; *Figure 3*).

Over-production of the DREB1A and CBF1/DREB1B cDNAs driven by the 35S CaMV promoter in transgenic plants significantly improved stress tolerance to drought and freezing. However, the DREB1A transgenic plants revealed severe growth retardation under normal growth conditions. The DREB1A cDNA driven by the stress-inducible *rd29A* promoter was expressed at low level under unstressed control conditions and strongly induced by dehydration, salt and cold stresses. The *rd29A* promoter minimized negative effects on growth of plants, whereas the 35S CaMV promoter caused severe growth retardation under normal growth conditions (M. Kasuga, Q. Liu, S. Miura, K. Yamaguchi-Shinozaki and K. Shinozaki, unpublished data). Moreover, this stress-inducible promoter enhanced tolerance to drought, salt and freezing at higher levels than that of the 35S CaMV promoter.

5. Signal perception and signal transduction in drought stress

Signal-transduction pathways from the sensing of dehydration signal or osmotic change to the expression of various genes, and the signalling molecules that function in stress signalling have not been extensively studied in plants. Signal-transduction pathways in drought stress response have been studied based on the knowledge in yeast and animal systems (*Figure 4*).

A two-component system functions in sensing osmotic stress in bacteria and yeast (Wurgler-Murphy and Saito, 1997). Plants as well as cyanobacteria contain many genes encoding sensor histidine kinases and response regulator homologues, which suggests the involvement of similar osmosensing mechanisms in higher plants. Recently, we isolated an *Arabidopsis* cDNA (ATHK1) encoding the two component histidine kinase, a yeast osmosensor Sln1 homologue, by PCR. ATHK1 has a typical histidine kinase domain and a receiver domain like Sln1, and has different structure in the N-terminal domain from that of ETR1 or an ethylene receptor (T. Urao, K. Yamaguchi-Shinozaki, T. Hirayama and K. Shinozaki, unpublished data). Over-expression of cATHK1 suppressed the lethality of a temperature-sensitive osmosensing-defective yeast *sln1* mutant (*sln1-ts*). The ATHK1 transcript was more abundant in roots than other tissues under normal growth conditions and accumulated in conditions of high salinity and low temperature. These results suggest that ATHK1 might function in signal perception during drought stress in *Arabidopsis*. Recently, we have cloned four cDNAs encoding a response regulator (Urao *et al.*, 1998), and three cDNAs encoding a phospho-relay intermediator containing an HPt domain (Miyata *et al.*, 1998) from *Arabidopsis*. These observations suggest that plants have an osmosensing and signalling system similar to that of yeast. Of course, other sensing mechanisms may function during drought stress responses, such as mechanical sensors of cytoskeletons and sensors of superoxides produced by stress (*Figure 4*).

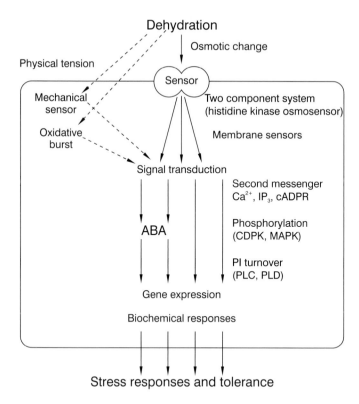

Figure 4. Second messengers and factors involved in the signal perception and the signal transduction in drought stress response. Two-component histidine kinase is thought to function as an osmosensor in plants. Ca^{2+} and inositol-3-phosphate (IP_3) are most probable second messengers of the dehydration signal. The phosphorylation process functions in water stress and abscisic acid (ABA) signal-transduction pathways. Phosphatidylinositol (PI) turnover is also involved in drought stress response. ABA plays important roles in the regulation of gene expression as well as physiological responses during water stress. Several ABA signal-transduction pathways are reported. CDPK, calcium-dependent protein kinase; MAPK, mitogen-activated protein kinase; PLC, phospholipase C; PLD, phospholipase D.

MAPK is involved in the signal-transduction pathways associated not only with growth factor-dependent cell proliferation but also with environmental stress responses in yeast and animals. Many cDNAs for proteins involved in MAPK cascade, MAPK, MAPKK, MAPKKK have been cloned in plants (Mizoguchi *et al.*, 1997). We have isolated more than nine genes for MAPK in *Arabidopsis*. There are at least four sub-families of MAPK based on phylogenetic analysis. One of the *Arabidopsis* MAPK genes, ATMPK3, is induced at the mRNA level by drought, low temperature, high salinity and touch (Mizoguchi *et al.*, 1996). Moreover, a gene for MAPKKK (ATMEKK1) is induced by similar stresses. We demonstrated rapid and transient activation of MAPK activities of ATMPK4 in *Arabidopsis* plants by low temperature, humidity change, wounding and touch (K. Ichimura, T. Mizoguchi and K. Shinozaki, unpublished data). Recently, alfalfa MAPK, MMK4, was demonstrated to be activated at post-translational levels by a variety of stresses including drought, low temperature and mechanical stim-uli (Jonak *et al.*, 1996). The MMK4 gene is also induced by these stresses at the tran-scriptional level. These observations indicate that certain MAPK cascade might function in the signal-transduction pathways in drought stress response.

Stomata closure is well characterized as a model system in the responses of plant cells to dehydration stress and ABA treatment (Shinozaki and Yamaguchi-Shinozaki, 1997). During stomata closure, the level of cytoplasmic Ca^{2+} is increased, which sug-gests that Ca^{2+} functions as a second messenger in the osmotic stress response. In animal cells, inositol-3-phosphate (IP_3) is involved in the release of Ca^{2+} into the cytoplasm from intracellular stores, and it may play a similar role in plant cells. Ca^{2+} and IP_3 are the most probable candidates as second messengers in drought stress responses in plant cells. Two cDNAs for calcium-dependent protein kinases (CDPKs) were isolated from dehydrated *Arabidopsis* (Urao *et al.*, 1994). Moreover, a gene for phospholipase C (PLC) is rapidly induced by drought and salt stresses in *Arabidopsis* (Hirayama *et al.*, 1995). Recently, a gene for phosphatidylinositol-4-phosphate-5-kinase (PIP5K) was demonstrated to be induced by water stress (Mikami *et al.*, 1998). The stress-inducible CDPKs, PLC and PIP5K might function in the signal-transduction cascade under water stress. In animals, PIP5K catalyses the production of phosphatidylinositol-4,5-bisphos-phate (PIP_2) from phosphatidylinositol-4-phosphate. PLC digests PIP_2 to generate two second messengers, IP_3 and diacylglycerol. IP_3 induces the release of Ca^{2+} into the cyto-plasm, which in turn causes various responses in the cytoplasm. In plants, a similar sys-tem may function in water stress response.

ABA plays important roles in drought stress responses. ABA is involved in not only stomata closure but also induction of many genes. Several mutants in ABA signalling have been identified and their genes encode protein phosphatases (*abi1* and *abi2*) and farnesyl transferase (*era1*) (Bonetta and McCourt, 1998). These suggest that protein dephosphorylation and protein farnesylation are involved in ABA signalling. However, various signalling molecules seem to be involved in ABA signalling, such as phospha-tidic acid and cyclic ADP ribose. Various protein kinases and enzymes involved in phos-pholipid metabolism have been reported in plants and are thought to function in signal-transduction pathways including drought stress and ABA responses. MAPK cas-cades and CDPK were suggested to be involved in drought stress response and ABA signalling. Complex signalling cascades are thought to function in molecular responses to drought stress. Molecular analyses of the signalling process is in progress based on genetics and gene cloning.

6. Key questions

Details of molecular mechanisms in drought stress response and tolerance are uneluci-dated. These include the sensing mechanisms of osmotic stress, modulation of the stress signals to cellular signals, transduction of the cellular signals to the nucleus, second messengers involved in stress signal transduction, roles of ABA in signalling process, transcriptional control of stress-inducible genes, and the function and co-operation of stress-inducible genes allowing drought stress tolerance.

Molecular mechanisms of drought stress response and tolerance have been actively studied during the last 10 years (Shinozaki and Yamaguchi-Shinozaki, 1997). Many genes that are regulated by drought stress have been reported in a variety of plants. Analyses of stress-inducible gene expression have revealed the presence of multiple signal-transduction pathways between the perception of drought stress signal and gene expression. At least four different transcription systems have been suggested to function in the regulation of dehydration-inducible genes; two are ABA-responsive and two are ABA-independent (Shinozaki and Yamaguchi-Shinozaki, 1997). This explains the complex stress response observed after exposure of plants to drought stress. Some genes are rapidly induced by drought stress within 10 min whereas others are slowly induced in a few hours after the accumulation of endogenous ABA. Several genes for transcription factors are induced by drought stress and ABA at transcriptional levels, which may be involved in the regulation of slowly expressed genes whose products function in stress tolerance and adaptation. In addition, many genes for factors involved in the signal-transduction cascades, such as protein kinases and enzymes involved in PI turnover, are up-regulated by drought stress signals (Shinozaki and Yamaguchi-Shinozaki, 1997). These signalling factors might be involved in the amplification of the stress signals and adaptation of plant cells to drought stress conditions.

However, no direct evidence has been obtained on the functions of these signalling molecules. Molecules that function as osmosensors and ABA receptors have not been identified. Based on the knowledge of osmosensors in yeast and bacteria, cloning of homologues of the two-component histidine kinase as an osmosensor is in progress in higher plants. Molecular analyses of these factors should provide a better understanding of the signal-transduction cascades during drought stress. Transgenic plants that modify the expression of these genes will give more information on the function of their gene products. Complex mechanisms seem to be involved in gene expression and signal transduction in response to drought stress. Genetic analyses of drought-resistant or drought-sensitive mutants are still difficult but have to be extensively performed with novel strategies. *Cis-* and *trans-*acting elements involved in stress-responsive gene expression have extensively studied. However, details of molecular mechanisms of regulating plant genes by drought stress remain to be solved.

7. Future directions

Genetic analysis of *Arabidopsis* mutants with rd29A drought-inducible promoter: luciferase reporter gene suggests complex signalling pathways in drought, salt and cold stress responses (Ishitani *et al.*, 1997). It is expected that genetic analyses of these mutants as well as drought-resistant or drought-sensitive mutants will provide more information on signal transduction in response to drought stress. Sequencing of the *Arabidopsis* genome is now in progress and will be completed by the year 2001, which means that structures of all the 20 000 genes of *Arabidopsis* will be determined in a few years. All the stress-inducible genes will be identified by systematic analysis of gene expression. In the next decade, we think it will be important to develop novel methods to analyse complex networks of stress responses of higher plants. Construction of insertion mutant lines using T-DNA and transposons will become more important to analyse functions of disrupted genes in plants. Reverse genetic approach as well as classical genetics will become more important to understand not only functions of stress-inducible genes but also complex signalling processes in environmental stress responses. Efficient gene disruption methods as well as transgenic approaches using antisense or sense constructs will also contribute to the precise understanding of molecular mechanisms of stress response.

Acknowledgements

This work was supported by the Program for Promotion of Basic Research Activities for Innovative Bioscience, the Special Co-ordination Fund of the Science and Technology Agency of the Japanese Government, a Grant-in-Aid from the Ministry of Education, Science and Culture of Japan and Human Frontier Science Program.

References

Abe, H., Yamaguchi-Shinozaki, K., Urao, T., Iwasaki, T., Hosokawa, D. and Shinozaki, K. (1997) Role of *Arabidopsis* MYC and MYB homologs in drought- and abscisic acid-regulated gene expression. *Plant Cell* **9:** 1859–1868.

Bonetta, D. and McCourt, P. (1998) Genetic analysis of ABA signal transduction pathways. *Trends Plant Sci.* **6:** 231–235.

Hirayama, T., Ohto, C., Mizoguchi, T. and Shinozaki, K. (1995) A gene encoding a phosphatidyl-inositol-specific phospholipase C is induced by dehydration and salt stress in *Arabidopsis thaliana*. *Proc. Natl Acad. Sci. USA* **92:** 3903–3907.

Ishitani, M., Xiong, L., Stevenson, B. and Zhu, J.K. (1997) Genetic analysis of osmotic and cold stress signal transduction in *Arabidopsis*: interactions and convergence of abscisic acid-dependent and abscisic acid-independent pathways. *Plant Cell* **9:** 1–16.

Jaglo-Ottosen, K.R., Gilmour, S.J., Zarka, D.G., Schabenberger, O. and Thomashow, M.F. (1998) Arabidopsis CBF1 overexpression induces *coe* genes and enhances freezing tolerance. *Science* **280:** 104–106.

Jonak, C., Kiegerl, M., Ligterink, W., Barker, P.J., Huskisson, N.S. and Hirt, H. (1996) Stress signaling in plants: a MAP kinase pathway is activated by cold and drought. *Proc. Natl Acad. Sci. USA* **93:** 11274–11279.

Kiyosue, T., Yamaguchi-Shinozaki, K. and Shinozaki, K. (1993) Characterization of cDNA for a dehydration-inducible gene that encodes a Clp A, B like protein in *Arabidopsis thaliana*. *Biochem. Biophys. Res. Commun.* **196:** 1214–1220.

Koizumi, M., Yamaguchi-Shinozaki, K., Tsuji, H. and Shinozaki, K. (1993) Structure and expression of two genes that encode distinct drought-inducible cysteine proteases in *Arabidopsis thaliana*. *Gene* **129:** 175–182.

Liu, Q., Sakuma, Y., Abe, H., Kasuga, M., Miura, S., Yamaguchi-Shinozaki, K. and Shinozaki, K. (1998) Two transcription factors, DREB1 and DREB2, with an EREBP/AP2 DNA binding domain, separate two cellular signal transduction pathways in drought- and low temperature-responsive gene expression, respectively, in *Arabidopsis*. *Plant Cell* **10:** 1391–1406.

Mikami, K., Katagiri, T., Iuchi, S., Yamaguchi-Shinozaki, K. and Shinozaki, K. (1998) A gene encoding phosphatidylinositol-4-phosphate 5-kinase is induced by water stress and abscisic acid in *Arabidopsis thaliana*. *Plant J.* **15:** 563–568.

Miyata, S., Urao, T., Yamaguchi-Shinozaki, K. and Shinozaki, K. (1998) Characterization of genes for two component phospho-relay mediators with a single HPt domain in *Arabidopsis thaliana*. *FEBS Lett.* **437:** 11–14.

Mizoguchi, T., Irie, K., Hirayama, T., Hayashida, N., Yamaguchi-Shinozaki, K., Matsumoto, K. and Shinozaki, K. (1996) A gene encoding a MAP kinase kinase is induced simultaneously with genes for a MAP kinase and an S6 kinase by touch, cold and water stress in *Arabidopsis thaliana*. *Proc. Natl Acad. Sci. USA* **93**: 765–769.

Mizoguchi, T., Ichimura, K. and Shinozaki, K. (1997) Environmental stress response in plants: the role of mitogen-activated protein kinases (MAPKs). *Trends Biotechnol.* **15**: 15–19.

Nakashima, K., Kiyosue, T., Yamaguchi-Shinozaki, K. and Shinozaki, K. (1997) A nuclear gene, *erd1*, encoding a chloroplast-targeted Clp protease regulatory subunit homolog is not only induced by water stress but also developmentally up-regulated during senescence in *Arabidopsis thaliana*. *Plant J.* **12**: 851–861.

Shinozaki, K. and Yamaguchi-Shinozaki, K. (1997) Gene expression and signal transduction in water-stress response. *Plant Physiol.* **115**: 327–334.

Shinwari, Z.K., Nakashima, K., Miura, S., Kasuga, M., Seki, M., Yamaguchi-Shinozaki, K. and Shinozaki, K. (1998) An *Arabidopsis* gene family encoding DRE/CRT binding proteins involved in low-temperature-responsive gene expression. *Biochem. Biophys. Res. Commun.* **250**: 161–170.

Stockinger, E.J., Gilmour, S.J. and Thomashow, M.F. (1997) *Arabidopsis thaliana* CBF1 encodes an AP2 domain-containing transcription activator that binds to the C-repeat/DRE, a *cis*-acting DNA regulatory element that stimulates transcription in response to low temperature and water deficit. *Proc. Natl Acad. Sci. USA* **94**: 1035–1040.

Urao, T., Yamaguchi-Shinozaki, K., Urao, S. and Shinozaki, K. (1993) An Arabidopsis *myb* homolog is induced by dehydration stress and its gene product binds to the conserved MYB recognition sequence. *Plant Cell* **5**: 1529–1539.

Urao, T., Katagiri, T., Mizoguchi, T., Yamaguchi-Shinozaki, K., Hayashida, N. and Shinozaki, K. (1994) Two genes that encode Ca^{2+}-dependent protein kinases are induced by drought and high-salt stresses in *Arabidopsis thaliana*. *Mol. Gen. Genet.* **224**: 331–340.

Urao, T., Yakubov, B., Yamaguchi-Shinozaki, K. and Shinozaki, K. (1998) Stress-responsive expression of genes for two-component response regulator-like proteins in *Arabidopsis thaliana*. *FEBS Lett.* **427**: 175–178.

Wurgler-Murphy, S.M. and Saito, S. (1997) Two-component signal transducers and MAPK cascades. *Trends Biochem. Sci.* **22**: 172–176.

Yamaguchi-Shinozaki, K. and Shinozaki, K. (1993) The plant hormone abscisic acid mediates the drought-induced expression but not the seed-specific expression of *rd22*, a gene responsive to dehydration stress in *Arabidopsis thaliana*. *Mol. Gen. Genet.* **238**: 17–25.

Yamaguchi-Shinozaki, K. and Shinozaki, K. (1994) A novel *cis*-acting element in an *Arabidopsis* gene is involved in responsiveness to drought, low-temperature, or high-salt stress. *Plant Cell* **6**: 251–264.

Chapter 18

Molecular characterization of two chloroplastic proteins induced by water deficit in *Solanum tuberosum* L. plants: involvement in the response to oxidative stress

Pascal Rey, Ghislaine Pruvot, Benjamin Gillet and Gilles Peltier

1. Background

Land plants are frequently subjected to periods of water deficit that affect productivity. Upon moderate water stress conditions, photosynthetic CO_2 assimilation is reduced due to an increased resistance to CO_2 diffusion consecutive to stomatal closure. More severe stress conditions provoke a significant, but reversible decrease in photosynthetic CO_2 fixation due to an inhibition of Calvin cycle enzymes (Cornic *et al.*, 1992; Kaiser, 1987). Photosynthetic electron transfer reactions, that take place in thylakoid membranes, have been shown to be remarkably resistant to dehydration stress since irreversible inhibition in electron transfer capacity occurs only upon drastic water deficit conditions, that is for leaf relative water contents (RWC) lower than 40% (Kaiser, 1987). As limitation of CO_2 fixation precedes inactivation of electron transfer reactions, an excess of reducing power is generated in chloroplasts of water-stressed plants. This excess results in the formation of active oxygen species, for instance at the photosystem I acceptor site (Smirnoff, 1993), causing cellular damage such as lipid peroxidation or protein modification. To prevent active oxygen formation, plants have evolved protective mechanisms such as heat dissipation of excess energy through carotenoids. Further, chloroplastic antioxidants and enzymes allow scavenging of active oxygen (Smirnoff, 1993). Up-regulation of superoxide dismutase (SOD) is likely to participate in dehydration tolerance (Smirnoff, 1993) and alfalfa plants over-expressing Mn-SOD within chloroplasts have been shown to exhibit higher photosynthetic efficiency during water deficit (McKersie *et al.*, 1996).

Whether higher plants have developed other mechanisms within the chloroplast conferring tolerance to the oxidative and osmotic stress conditions resulting from water deficit remains unknown. Among the numerous drought-induced genes that have been proposed to promote cellular tolerance of dehydration, most encode proteins localized

Plant Responses to Environmental Stress, edited by M.F. Smallwood, C.M. Calvert and D.J. Bowles.

in the cytosol (Ingram and Bartels, 1996). Until now, very few drought-induced chloroplastic proteins have been identified. High levels of betaine aldehyde dehydrogenase, that catalyses formation of glycine betaine, a compatible solute, have been reported in chloroplasts of sugar beet plants upon osmotic stress (McCue and Hanson, 1992). This solute has been also shown to stabilize photosystem II activity *in vitro* (Papageorgiou and Murata, 1995). In *Craterostigma plantagineum*, a resurrection plant able to survive extreme dehydration, three genes preferentially expressed upon water stress were found to encode chloroplastic desiccation stress proteins, DSP, that have been proposed to fulfil a protective function of chloroplast structures (Bartels *et al.*, 1992; Schneider *et al.*, 1993). Immunological studies revealed that two proteins, DSP 22, an ELIP-like protein, and DSP 34, were located in thylakoids whereas the third DSP 21, weakly homologous to a late embryogenesis abundant (LEA) protein, was found in stroma.

We have identified two chloroplastic proteins of 32 and 34 kDa exhibiting a highly enhanced synthesis in response to water deficit in plants of a cultivated species, *Solanum tuberosum*. These proteins were termed CDSP 32 and CDSP 34 for chloroplastic drought-induced stress protein (Pruvot *et al.*, 1996a, 1996b). Here, we present molecular characteristics of CDSPs, and hypothetical functions of the two proteins are discussed in relation to their sequence homologies and synthesis conditions.

2. CDSP 32 characterization

2.1 Molecular characteristics of CDSP 32

By analysing two-dimensional patterns of chloroplastic proteins, a stromal 32-kDa protein termed CDSP 32 was found to be highly synthesized in *Solanum tuberosum* plants subjected to water deficit (*Figure 1*). A cDNA encoding CDSP 32 was cloned by screening an expression library with a serum raised against protein N-terminus. The cDNA

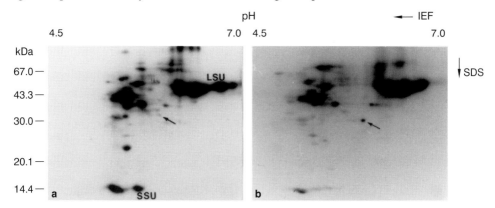

Figure 1. Two-dimensional patterns of labelled stromal proteins from control and drought-stressed *Solanum tuberosum* plants. *In vivo* labelling using L-[^{35}S]methionine was performed on leaflets from control plants (95% RWC, a) and drought-stressed plants (65% RWC, b). Stromal proteins were separated by two-dimensional electrophoresis and subjected to fluorography. The arrow points to the CDSP 32 protein. LSU and SSU, large and small subunits of Rubisco; IEF, isoelectric focusing; SDS, sodium dodecyl sulphate.

(1149 bp, EMBL Y09987) encodes a 296-residue precursor with a chloroplastic transit peptide of 53 amino acids (Rey *et al.*, 1998). Mature CDSP 32 (243 amino acids) shares, in its C-terminal region, 34.2% identity and 50.4% similarity in 111 residues overlap with tobacco thioredoxin h$_2$ (SWISS-PROT Q07090). Thioredoxins are ubiquitous proteins containing an active site CGPC conferring disulphide reductase activity (Holmgren, 1985). Significantly, CDSP 32 contains a CGPC motif and other well-conserved amino acids recognized as important for structure and activity in *Escherichia coli* thioredoxin (*Figure 2*). Sequence analysis also revealed noticeable homology between CDSP 32 N-terminus (121 residues) and thioredoxin h from *Arabidopsis thaliana* (SWISS-PROT P29448) and alignment of CDSP 32 N- and C-terminal half parts revealed 28.3% identity and 46.0% similarity. No CGPC-like sequence is found in the N-terminal region, but strikingly, other residues well-conserved in thioredoxins are present in this region (Rey *et al.*, 1998), suggesting that *CDSP 32* gene arose from duplication of a thioredoxin-type gene.

We investigated whether CDSP 32 possesses thioredoxin activity by producing a recombinant protein corresponding to CDSP 32 C-terminal 122 residues and by measuring the ability of the protein to catalyse reduction of insulin disulphide bridges in the presence of dithiothreitol (Holmgren, 1979). The recombinant protein was found to reduce insulin (Rey *et al.*, 1998), indicating that CDSP 32 displays thioredoxin activity in its C-terminal region.

In other respects, accumulations of both CDSP 32-related transcript and protein were observed in tomato plants subjected to water stress (Rey *et al.*, 1998) and we isolated a cDNA encoding a protein highly similar to CDSP 32 in *Arabidopsis thaliana* (P. Rey, N. Becuwe and F. Eymery, unpublished data). These data suggest that CDSP 32-related proteins are probably present in many plant species.

```
CDSP 32     128  VVQLHSREDVEKVIQDHKIDKKLIVLDVGLKHCGPCVKVYPTVIKLSKQMADTVVFARM
THII_TOBAC    7  VIGVHTVDAWNEHLQKGIDDKKLIVVDFTASWCGPCKFIASFYAELAKKMP-TVTFLKV
THIH_TOBAC   14  VFGCHKVEEWNEYFKKGVETKKLVVVDFTASWCGPCRFIAPILADIAKKMP-HVIFLKV
THIF_SPIOL   85  VTEVNKDTFWPIVKAAGDKP---VVLDMFTQWCGPCKAMAPKYEKLAEEYL-DVIFLKL
THIM_SPIOL    8  EVQDVNDSGWKEFVLQSSEP---SMVDFWAPWCGPCKLIAPVIDELAKEYSGKIAVTKL
THIO_ECOLI    3  KIIHLTDDSFDTDVLKADGA---ILVDFWAEWCGPCKMIAPILDEIADEYQGKLTVAKL
Consensus                              V VDF A  WCGPC     P    L              K

CDSP 32     187  NGDENDSCMQFLKDMDVIEVPTFLFIRDGEICGRYVGSG-KGELIGEILRYQGVRVTY-
THII_TOBAC   65  DVDELKS---VATDWAVEAMPTFMFLKEGKIVDKVVGAK-KDELQQTIAKHIS-STSTA
THIH_TOBAC   72  DVDELKT---VSAEWSVEAMPTFVFIKDGKEVDRVVGAK-KEELQQTIVKHAAPATVTA
THIF_SPIOL  140  --DCNQENKTLAKELGIRVVPTFKILKENSVVGEVTGAKYDKLLEAIQAARSS------
THIM_SPIOL   64  NTDEAPG---IATQYNIRSIPTVLFFKNGERKESIIGDVSKYQL--------------
THIO_ECOLI   59  NIDQNPG---TAPKYGIRGIPTLLLFKNGEVAATKVGALSKGQLKEFLDANLA- ----
Consensus          N D N     A    PT  FK G   VGA  K  L
```

Figure 2. Multiple sequence alignment of CDSP 32 C-terminal region, plant and *Escherichia coli* thioredoxins. Thioredoxin consensus amino acids (Eklund *et al.*, 1991) and conserved residues are in bold. Underlined amino acids correspond to structurally important residues deduced from three-dimensional models. Plant and *E. coli* thioredoxin sequence names correspond to SWISS-PROT identification. THII_TOBAC, tobacco thioredoxin h$_2$; THIH_TOBAC, tobacco thioredoxin h$_1$; THIF_SPIOL, spinach thioredoxin f; THIM_SPIOL, spinach thioredoxin m; THIO_ECOLI, *E. coli* thioredoxin.

2.2 Expression of the CDSP 32 gene

Low levels of CDSP 32 mRNA were detected in leaves from control potato plants and from plants subjected to mild water deficit (84% RWC). The transcript abundance was much higher when the RWC reached 70% and *in vivo* labelling revealed a substantial increase in CDSP 32 synthesis upon similar stress conditions (Rey *et al.*, 1998). When severely wilted plants were re-watered for 16 h, strong decreases in both transcript and protein abundances were noticed (Rey et al., 1998). CDSP 32 accumulation has been also observed in potato plants subjected to high salt treatment, but not in plants exposed to low temperatures (Pruvot *et al.*, 1996b).

   We investigated the role of abscisic acid (ABA), a phytohormone mediating gene expression upon osmotic stress, in the regulation of *CDSP 32* expression. Exogenously applied ABA was found not to induce CDSP 32 accumulation in potato plants (Pruvot *et al.*, 1996b). *CDSP 32* expression was also examined in *flacca* tomato mutants impaired in ABA biosynthesis and we noticed that water stress induced substantial accumulations of CDSP 32 transcript and protein in both wild-type and *flacca* tomato plants (Rey *et al.*, 1998). These data indicate that ABA is probably not required in the regulation of CDSP 32 synthesis.

2.3 Role of CDSP 32

In potato plants, CDSP 32 is highly synthesized upon severe water stress conditions (Rey *et al.*, 1998) and high salinity stress (Pruvot *et al.*, 1996b). Upon such conditions, the part of non-stomatal inhibition of photosynthesis turns significant due to chemical modifications resulting from high solute concentration (Kaiser, 1987) and change in redox state (Smirnoff, 1993). These modifications are likely to provoke enzyme inactivation (Kaiser, 1987) due to protein unfolding or aggregation through intermolecular disulphide bonds (Levitt, 1980). Preservation of enzyme structure appears to be critical upon severe stress conditions and, in animal cells, thioredoxin has been shown to regenerate the activity of proteins damaged by oxidative stress (Fernando *et al.*, 1992). Based on the putative function deduced from sequence and activity characteristics, we propose that CDSP 32 may play a role in preservation of protein native structure by reducing intermolecular disulphide bonds. Further investigations will be needed to determine the catalytic properties of CDSP 32, identify its substrate(s) and assess its function in relation to changes in thiol:disulphide redox potential occurring during water deficit.

3. CDSP 34 characterization

3.1 Molecular characteristics of CDSP 34

The accumulation of CDSP 34, a 34-kDa thylakoid protein, was observed in potato plants subjected to water shortage (*Figure 3*). Western analysis of thylakoid subfractions (Pruvot *et al.*, 1996a) and immunocytolocalization (Eymery and Rey, 1999), using a serum raised against CDSP 34 N-terminus, indicated a preferential localization of the protein in stromal lamellae thylakoids and in their extrinsic fraction. A PCR product,

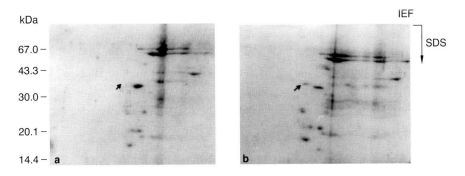

Figure 3. Two-dimensional patterns of thylakoid proteins from control and drought-stressed *Solanum tuberosum* plants. Purified chloroplasts were prepared from control (95% RWC, a) and drought-stressed (60% RWC, b) plants. Thylakoid proteins were separated using two-dimensional electrophoresis (60 μg per gel) and revealed with coomassie blue. The arrow points to the CDSP 34 protein.

amplified from potato genomic DNA with primers deduced from CDSP 34 microsequences, was used to clone a CDSP 34 cDNA (Gillet *et al.*, 1998). The full-length cDNA (1261 bp, EMBL Y15269) contains an open reading frame encoding a 326-residue precursor with a chloroplastic transit peptide of 63 amino acids. Sequence analysis revealed high similarities between CDSP 34 and two chromoplastic proteins, fibrillin from *Capsicum annuum* (Deruère *et al.*, 1994) and CHRC (for <u>ch</u>romoplast protein <u>C</u>) from *Cucumis sativus* (Vishnevetsky *et al.*, 1996). Mature CDSP 34 and fibrillin contain 263 amino acids and display 90.8% identity and 95.0% similarity (*Figure 4*). Mature CHRC shares 73.7% identity and 82.2% similarity with CDSP 34. Fibrillin is the major protein component of carotenoid-accumulating structures, termed fibrils, in ripening pepper fruit (Deruère *et al.*, 1994) and CHRC is a specific chromoplast carotenoid-associated protein accumulating in cucumber corollas (Vishnevetsky *et al.*, 1996). CDSP 34, fibrillin and CHRC appear to belong to a group of well-conserved proteins expressed in dicotyledonous and monocotyledonous plants (*Figure 4*). We recently obtained immunological data showing that water deficit induced accumulation of CDSP 34-related proteins in different species such as *Nicotiana tabacum* and *Arabidopsis thaliana* (P. Rey and S. Cuiné, unpublished data).

3.2 Expression of the CDSP 34 gene

A strong increase in the CDSP 34 transcript level was observed from mild water deficit conditions (84% RWC). *In vivo* labelling experiments and Western analyses showed enhanced CDSP 34 synthesis and protein accumulation from similar stress conditions. Re-watering of wilted plants (65% RWC) resulted in severe decreases in CDSP 34 mRNA level and protein synthesis, but no noticeable change in CDSP 34 abundance was observed, indicating a relatively low turnover of the protein (Gillet *et al.*, 1998; Pruvot *et al.*, 1996a). CDSP 34 accumulation was also observed in potato plants exposed to low temperatures, grown upon high salt conditions or subjected to high illumination (Gillet *et al.*, 1998; Pruvot *et al.*, 1996b).

```
CDSP 34    64 ATNYDKEDEWGPEVEQI----RPGGVAVVEEEPPKEPSEIELLKKQLADSLYGTNRGLSASSETRAEIVE
Fibrillin  60 ATNYDKEDEWGPELEQI----NPGGVAVVEEEPPKEPSEMEKLKKQLTDSFYGTNRGLSASSETRAEIVE
CHRC       59 AVLND--DEWGEDKDEKYGDDSSVAVAEKEEEKPLEPSEIYKLKKALVDSFYGTDRGLRVSRDTRAEIVE
Maize       3                                                          RASTEVRGEVVE
Synech.     1                                   MPNSMDANMDFKTNLLEAIAGKNRGLLASDRDRVAILS


CDSP 34   130 LITQLESKNPNPAPTEALTLLNGKWILAYTSFSSLFPLLSRGNLPLVRVEEISQTIDSESFTVQNSVVFA
Fibrillin 126 LITQLESKNPTPAPTEALSLLNGKWILAYTSFSSLFPLLARGNLLPVRVEEISQTIDAETLTVQNSVVFA
CHRC      127 LITQLESKNPTPAPTEALTLLNGKWILAYTTFASLFPLLSR-NLPLVKVEEISQTIDSENLTVQNSVQFS
Maize      38 LVTQLEALNPTTAPVESPDLLDGNWILIYTAYSELLPILAAGATPLVKVKQISQEIDSKIMTIXNASTLT
Synech.    39 AVEKLEDYNPHPKPLQEKNLLDGNWRLLYTSSQSILGL---NRLPLLQLGQIYQYIDVAGSRVVNLAEIE


CDSP 34   200 G-PLATTSISTNAKFEVRSPKRVQIKFEEGIIGTPQLTDSIVLPENVEFLGQKIDLSPFKGLITS-VQDT
Fibrillin 196 G-PLSTTSISTNAKFEVRSPKRLQINFEEGIIGTPQLTDSIELPENVEFLGQKIDLSPFKGLITS-VQDT
CHRC      196 G-PLATTSITTNAKFEVRSPLRVHIKFEEGVIGTPQLTDSIVIPDNVDFLGQKIDFTPFNGIISS-LQDT
B.rapa      1      ATTSISTNAKFEVRSPKRVQIKFEEGVIGTPQLTDSIEIPEYVEFLGQKIDLTPIRGLLTSGTKTR
Maize     248 T-PFASFSFSATASFEVQ
Synech.   106 GIPFLESLVSVVASFIPVSDKRIEVKFERSILGLQKILNY--------------QSPLKF----IQQI-


CDSP 34   268 ASSVAKSISSQPPIKFPITNNNAQSWLLTTYLDDELRISRGDAGSVFVLIKEGSPLLKP
Fibrillin 264 ATSVAKSISSQPPIKFPISNSYAQSWLLTTYLDAELRISRGDAGSIFVLIKEGSPLLKP
CHRC      264 ASNVAKTISSQPPIKFSISNTRVESWLLTTYLDEDLRISRGDGGSVFVLLKEGSSFLSL
B.rapa    198 ATSCGEPISSQPPVEILFCPGGAQSWLLTAYLDKDIRISRGDGGSVFVLIK
Synech.   156 --STGKRFL---PADFNLPGRDNAAWLEITYLDEDLRISRGNEGNVFILAKV
```

Figure 4. Multiple sequence alignment of CDSP 34 (EMBL Y15269), fibrillin (EMBL X71952), CHRC (EMBL X95593), protein sequences deduced from a *Brassica rapa* expressed sequence tag (EST) (GenBank BNAF0708E), a maize EST (GenBank AA072431) and a hypothetical protein termed 'fibrillin' from *Synechocystis* sp. strain PCC 6803 (Cyanobase sll1568). Identical residues are shown in bold.

ABA spraying caused an increase in CDSP 34 abundance in well-watered potato plants (Pruvot *et al.*, 1996b), suggesting a possible effect of the phytohormone in CDSP 34 synthesis regulation. In tomato plants subjected to water stress, similar accumulations of CDSP 34 transcript were noticed in wild-type and *flacca* ABA-deficient plants. However, whereas CDSP 34 accumulated in wild-type plants upon water stress, no increase in the protein level was observed in *flacca* mutants. CDSP 34 accumulation in *flacca* plants was restored by ABA spraying (Gillet *et al.*, 1998). These data lead us to propose that ABA might act at a post-transcriptional level in the regulation of CDSP 34 synthesis.

3.3 Role of CDSP 34

Fibrillin and CHRC were originally detected in chromoplastic carotenoid-accumulating structures. Our data clearly indicate that fibrillin-related proteins also accumulate within chloroplasts in many species upon water stress. Fibrillin has been suggested to structurally stabilize, through association with lipids, carotenoid-accumulating structures (Deruère *et al.*, 1994). Pozueta-Romero *et al.* (1997) proposed that fibrillin, due to its association with plastoglobuli, might also play a role in the storage of lipophilic compounds. Accordingly, we propose that CDSP 34 may participate in structural stabilization of stromal lamellae thylakoids and prevent damage resulting from environmental constraints. The analysis of transgenic plants modified in the expression of *CDSP 34* should provide information about the function of the protein upon stress conditions.

4. Involvement of the two proteins in the response to oxidative stress

As water deficit provokes the formation of active oxygen species due to an excess of excitation energy in relation to reduced CO_2 assimilation, we investigated whether the up-regulation of CDSPs synthesis upon water stress could result from change in the redox state within chloroplasts. We examined the expression levels of *CDSPs* genes in *Solanum tuberosum* plants subjected to (photo)-oxidative treatments such as exposition to high light (1400 μmol m^{-2} s^{-1}) and low temperature (8°C), spraying with methyl viologen (50 μM) and γ irradiation (100 Gray). Substantial increases in abundances of both proteins were observed upon the three treatments without any significant change in leaf relative water content (M. Broin and P. Rey, unpublished data), indicating the participation of signalling pathways linked to oxidative stress in the up-regulation of CDSPs synthesis. Taken together, the sequence characteristics of CDSPs and their synthesis conditions lead us to propose that the two proteins participate in mechanisms conferring tolerance within chloroplasts to the oxidative stress conditions resulting from water deficit.

Acknowledgements

The authors wish to thank N. Becuwe, A. Beyly, M. Broin, S. Cuiné, F. Eymery, J. Massimino and D. Rumeau for their valuable participation in CDSPs characterization.

References

Bartels, D., Hanke, C., Schneider, K., Michel, D. and Salamini, F. (1992) A desiccation-related *Elip*-like gene from the resurrection plant *Craterostigma plantagineum* is regulated by light and ABA. *EMBO J.* **11**: 2771–2778.

Cornic, G., Ghashghaie, J., Genty, B. and Briantais, J.M. (1992) Leaf photosynthesis is resistant to a mild drought stress. *Photosynthetica* **27**: 295–309.

Deruère, J., Römer, S., d'Harlingue, A., Backhaus R.A., Kuntz, M. and Camara, B. (1994) Fibril assembly and carotenoid overaccumulation in chromoplasts: a model for supramolecular lipoprotein structures. *Plant Cell* **6**: 119–133.

Eklund, H., Gleason, F.K. and Holmgren, A. (1991) Structural and functional relations among thioredoxins of different species. *Proteins: Structure Function Genet.* **11**: 13–28.

Eymery, F. and Rey, P. (1999) Immunocytolocalization of two chloroplastic drought-induced stress proteins in well-watered or wilted *Solanum tuberosum* L. plants. *Plant Physiol. Biochem.* **37** (in press).

Fernando, M.R., Nanri, H., Yoshitake, S., Nagata-Kuno, K. and Minakami, S. (1992) Thioredoxin regenerates proteins inactivated by oxidative stress in endothelial cells. *Eur. J. Biochem.* **209**: 917–922.

Gillet, B., Beyly, A., Peltier, G. and Rey, P. (1998) Molecular characterization of CDSP 34, a chloroplastic protein induced by drought stress in *Solanum tuberosum* L. plants and regulation of *CDSP 34* expression by ABA and high illumination. *Plant J.* **16**: 257–262.

Holmgren, A. (1979) Thioredoxin catalyzes the reduction of insulin disulfides by dithiothreitol and dihydrolipoamide. *J. Biol. Chem.* **254**: 9627–9632.

Holmgren, A. (1985) Thioredoxin. *Annu. Rev. Biochem.* **54**: 237–271.

Ingram, J. and Bartels, D. (1996) The molecular basis of dehydration tolerance in plants. *Annu. Rev. Plant Physiol. Plant Mol. Biol.* **47**: 377–403.

Kaiser, W.M. (1987) Effects of water deficit on photosynthetic capacity. *Physiol. Plant.* **71:** 142–149.

Levitt, J. (1980) Drought tolerance. In: *Responses of Plants to Environmental Stresses,* Vol. II (ed. T.T. Kozlowski). Academic Press: London, pp. 129–186.

McCue, K.F. and Hanson, A.D. (1992) Salt-inducible betaine dehydrogenase from sugar beet: cDNA cloning and expression. *Plant Mol. Biol.* **18:** 1–11.

McKersie, B.D., Bowley, S.R., Harjanto, E. and Leprince, O. (1996) Water-deficit tolerance and field performance of transgenic alfalfa overexpressing superoxide dismutase. *Plant Physiol.* **111:** 1177–1181.

Papageorgiou, G.C. and Murata, N. (1995) The unusually strong stabilizing effects of glycine betaine on the structure and function of the oxygen-evolving Photosystem II complex. *Photosynthesis Res.* **44:** 243–252.

Pozueta-Romero, J., Rafia, F., Houlné, G., Cheniclet, C., Carde, J.P., Schantz, M-L. and Schantz, R. (1997) A ubiquitous plant housekeeping gene, *PAP*, encodes a major protein component of bell pepper chromoplasts. *Plant Physiol.* **115:** 1185–1194.

Pruvot, G., Cuiné, S., Peltier, G. and Rey, P. (1996a) Characterization of a novel drought-induced 34-kDa protein located in the thylakoids of *Solanum tuberosum* L. plants. *Planta* **198:** 471–479.

Pruvot, G., Massimino, J., Peltier, G. and Rey, P. (1996b) Effects of low temperature, high salinity and exogenous ABA on the synthesis of two chloroplastic drought-induced proteins in *Solanum tuberosum*. *Physiol. Plant.* **97:** 123–131.

Rey, P., Pruvot, G., Becuwe, N., Eymery, F., Rumeau, D. and Peltier, G. (1998) A novel thioredoxin-like protein located in the chloroplast is induced by water deficit in *Solanum tuberosum* L. plants. *Plant J.* **13:** 97–107.

Schneider, K., Wells, B., Schmelzer, E., Salamini, F. and Bartels, D. (1993) Desiccation leads to the rapid accumulation of both cytosolic and chloroplastic proteins in the resurrection plant *Craterostigma plantagineum* Hochst. *Planta* **189:** 120–131.

Smirnoff, N. (1993) The role of active oxygen in the response of plants to water deficit and desiccation. *New Phytol.* **125:** 27–58.

Vishnevetsky, M., Ovadis, M., Itzhaki, H., Levy, M., Libal-Weksler, Y., Adam, Z. and Vainstein, A. (1996) Molecular cloning of a carotenoid-associated protein from *Cucumis sativus* corollas: homologous genes involved in carotenoid sequestration in chromoplasts. *Plant J.* **10:** 1111–1118.

Chapter 19

Late embryogenesis abundant (LEA) proteins: expression and regulation in the resurrection plant *Craterostigma plantagineum*

Dorothea Bartels

1. Background

LEA proteins were initially discovered as non-storage proteins which accumulate abundantly during late stages of cotton embryo development, hence the name LEA (late embryogenesis abundant) (Dure *et al.*, 1981). Meanwhile it has become evident that LEA proteins occur ubiquitously in seeds of higher plants. Expression studies revealed that LEA proteins are generally associated with plant cell dehydration which is not restricted to desiccation of seeds during development but may also be caused by environmental cues such as dehydration of vegetative tissues, low temperature, increased salt solutions or application of the plant growth regulator abscisic acid (ABA). Expression of LEA proteins is either developmentally or environmentally regulated. LEA proteins disappear during the first hours of germination or in response to stress relief.

Genes encoding LEA proteins are fairly diverse and are organized in gene families. LEA proteins have been divided into classes based on conserved amino acid sequence motifs (Baker *et al.,* 1988; Dure *et al.*, 1989). Sequence homologues of different LEA genes exist in many species, although the list is very incomplete. The best studied group of LEA genes and proteins are the dehydrins which have been reported from a broad range of species and which are characterized by a conserved lysine-rich 15 amino acid consensus sequence (EKKGIMDKIKEKLPG) (Close, 1996; Close *et al.*, 1993). Although different groups of LEA proteins are not similar on the amino acid level, LEA proteins are characterized by their biased amino acid composition (cysteine or tryptophan residues are very rare or do not occur), by their solubility in water and by their solubility after boiling. LEA proteins have been reported in most cellular compartments of the plant cell including nuclei and chloroplasts (Asghar *et al.*, 1994; Goday *et al.*, 1994; Schneider *et al.*, 1993).

Plant Responses to Environmental Stress, edited by M.F. Smallwood, C.M. Calvert and D.J. Bowles.
© 1999 BIOS Scientific Publishers, Oxford.

At least two groups of LEA proteins in cotton appear to be uniformly distributed among different cell types of the embryo and are found at high molar concentrations (Roberts *et al.*, 1993). Despite thorough molecular studies of LEA genes and proteins our knowledge of their fundamental biochemical role is scarce. Correlative studies and biochemical features of LEA proteins strongly support a protective role in the plant cell during dehydration.

Functions of LEA proteins can be addressed in different experimental approaches: either by *in vitro* studies trying to demonstrate the protective properties of LEA proteins or by *in vivo* studies using transgenic plants as tools. Transgenic plants over-expressing different types of LEA genes have been established. Some reports demonstrated positive effects of LEA transgenes during osmotic stress situations (Imai *et al.*, 1996; Xu *et al.*, 1996) whereas others did not show an effect of LEA genes (Iturriaga *et al.*, 1992). The negative results could have various reasons: firstly the over-expression of one LEA gene may not be sufficient to cause an effect as several different LEA proteins are usually expressed at the same time or, secondly, the analytical and physiological methods to determine the effect of LEA proteins were too crude.

The protective effect of LEA-like gene products is supported from studies of cold-induced genes. Constitutive expression of the *Arabidopsis COR15a* gene enhanced the freezing tolerance of both chloroplasts and protoplasts (Artus *et al.*, 1996). The mechanism by which the COR15a protein is protective is a decrease of freeze-induced lamellar to hexagonal II phase transitions, which occur at the inner chloroplast envelope as a result of freeze-induced dehydrations (Steponkus *et al.*, 1998). An obvious approach to address functions of LEA proteins is the isolation of mutants. This approach is unfortunately not very promising because LEA genes occur in gene families and several very similar copies exist in the genome for most classes and species. Therefore one defective gene copy is unlikely to affect the phenotype.

2. Regulation of expression

The high abundance and strictly regulated expression of LEA genes in embryos and the inducibility by external stimuli make LEA genes excellent candidate genes to analyse tissue-specific expression and to isolate promoter elements responsible for environmental and ABA responses. The functional dissection of LEA gene promoters has primarily been done in transient expression systems utilizing the fact that most LEA genes can be activated by ABA. This work has identified several *cis*-acting elements involved in ABA-induced gene expression (for a review see Leung and Giraudat, 1998). One group of promoter elements contain the G-box ACGT core motif and are known as ABREs (ABA-responsive elements). Multimerized but not single copies of ABREs can confer ABA responsiveness to minimal promoters (Skriver *et al.*, 1991). ABRE-like sequence motifs have been reported for many LEA genes, although only very few have been functionally tested. In fact promoter analysis of several LEA genes suggest that many ABREs do not function or are not sufficient for ABA induction (e.g. Busk *et al.*, 1997; Shen *et al.*, 1996). Several proteins have been reported which bind DNA sequences with ACGT core motifs and may thus act as transcriptional activators (Guiltinan *et al.*, 1990;

Kim *et al.*, 1997; Nakagawa *et al.*, 1996; Oeda *et al.*, 1991). Most of them belong to the bZIP-type proteins and contain a basic region involved in DNA binding and a leucine repeat involved in dimerization.

Besides the ABREs several different *cis*-acting elements have been identified (e.g. Nelson *et al.*, 1994) and in some cases two *cis*-elements are necessary for ABA-mediated gene activation (Shen *et al.*, 1996). Despite the strict regulation of LEA genes, promoter studies indicate a diversity of different DNA motifs acting individually or in combination in ABA-mediated LEA gene expression. The nature of putative transcription factors that bind to these sequences are largely unknown. It became apparent recently that LEA gene transcription is not always mediated by ABA, but ABA-independent activation pathways exist (Yamaguchi-Shinozaki *et al.*, 1995). Recently a *trans*-acting factor involved in mediating the drought response was cloned from *Arabidopsis* (Liu *et al.*, 1998).

The isolation of transcription factors which activate a set of LEA proteins is very important, as this will contribute to an altered stress tolerance (Jaglo-Ottensen *et al.*, 1998). Many studies on expression and regulation of expression have been carried out. The limitations are our understanding of structure and function. Biochemical and biophysical studies with purified LEA proteins will be one approach to advance our understanding of LEA proteins.

3. LEA genes and proteins in the resurrection plant *Craterostigma plantagineum*

Besides seeds the small group of resurrection plants shows extreme tolerance to dehydration and the plants are able to recover completely upon rehydration from severe dehydration. As a representative of the resurrection plants *Craterostigma plantagineum* has been extensively studied on the molecular level. The genetic programme involved in desiccation tolerance of vegetative tissues in the resurrection plant is similar to that of seeds, in that both accumulate LEA-like proteins and soluble sugars. LEA genes encoding diverse classes of LEA proteins have been intensively studied in *C. plantagineum* (Piatkowski *et al.*, 1990; Schneider *et al.*, 1993; Velasco *et al.*, 1998). LEA genes from *C. plantagineum* have the following features: genes share DNA motifs and amino acid sequences with embryogenesis abundant genes and proteins from other plant species, they encode predominantly hydrophilic proteins, they are inducible in vegetative tissues by dehydration and exogenous ABA treatment to a very abundant level.

Table 1 presents an overview of the different classes of LEA genes isolated as cDNA clones from dehydrated or ABA-treated leaves and callus of *C. plantagineum*. Different classes of LEA genes were defined by DNA sequence similarities (*Table 1*). Each LEA class is often organized in small gene families: for example the group defined by pcC 6–19 comprises at least four different members. The expression patterns were analysed for all cDNA clones listed in *Table 1*. It was commonly observed that the encoded transcripts were abundantly expressed in response to dehydration or ABA treatment in vegetative tissues. All transcripts accumulate during a time course treatment except for the novel gene pcC edi-9 which is transiently expressed in response to dehydration and

Table 1. LEA-like genes from *Craterostigma plantagineum*

cDNA clone from C. plantagineum	Homologous LEA genes[a]	Subcellular localization of protein[b]	References[c]
pcC 6–19	Group 2 *Lea* Cotton *LeaD11* (dehydrin)	Cytoplasm	1
pcC 27–45	Cotton *Lea14*	Cytoplasm	1
pcC 11–24	*Arabidopsis rd29*	Cytoplasm	2
pcC 34–62	Cotton *LeaD7*	n.d.[d]	3
pcC 163	Cotton *LeaD29*	n.d.	3
pcC 26	Cotton *LeaD113*	n.d.	3
pcC 76	*Brassica napus*		4
pcC 3–06	Group 3 *Lea* (wheat)	Chloroplast stroma	1
pcC 13–62	*Arabidopsis Est*	Chloroplast thylakoid membrane	1
pcC edi-9	No homology found	n.d.	5
pcC C2	*Le25*-homologue (*Arabidopsis*)	n.d.	6

[a]The homologies are referred to the different cotton LEA genes, if possible.
[b]For a reference see Schneider *et al.* (1993).
[c]The cDNA clones are described in the following references: 1 = Piatkowski *et al.*, 1990; 2 = Velasco *et al.*, 1998; 3 = Allamillo *et al.*, 1994; 4 = Bockel *et al.*, 1998; 5 = C. Bockel, M. Rodrigo and D. Bartels, unpublished data; 6 = A. Ditzer, H.H. Kirch and D. Bartels, unpublished data.
[d]n.d. = not done.

which is restricted to specific tissues. The expression data suggest that activation of LEA genes is co-ordinately regulated in *C. plantagineum*. Many LEA genes are activated in a defined time course during dehydration and thus the cell is able to accumulate a high concentration of LEA proteins which may be present in different cellular compartments and which may exert a related but a broad spectrum of protective functions. It is speculated that the co-ordinated accumulation of all the different LEA proteins is an essential component of desiccation tolerance in *C. plantagineum*. The first structural studies were performed with the recombinant dehydrin-type LEA protein 6–19 or dsp 16. These studies suggested a highly flexible monomeric protein with a very low content of secondary structure which is in support of the hypothesis that the protein may be involved in water binding and/or mimicking hydration (Lisse *et al.*, 1996).

4. Regulation of gene expression

In order to understand the co-ordinated transcriptional activation of different LEA genes promoter studies of selected genes (*CDeT6–19*, *CDeT 27–45*, *CDeT 11–24*) were carried out. Dot blot matrix comparisons did not reveal common motifs between the three promoters. Therefore deletion analyses were performed with all three promoters using the bacterial gene glucuronidase (GUS) as a reporter for promoter activities (*Table 2*). *C. plantagineum* was not suitable for efficient stable transformation, therefore

Table 2. Glucuronidase activity as a measure for promoter activity of *Craterostigma plantagineum* LEA genes

Gene	Transient expression in *C. plantagineum*	Transgenic tobacco plants	Transgenic *Arabidopsis* plants	*Arabidopsis* promoter-*GUS* X EN35S-ABI-3[a]
CDeT27–45[a]	ABA inducibility on two promoter regions	Seeds (embryos) and pollen, no activity in leaves, not inducible by ABA and dehydration in leaves	Seeds In ABA-treated seedlings	Responsiveness to ABA Regained in leaves
CDeT6–19[a]	ABA inducibility on two promoter regions	Seeds (embryos) pollen, leaves ABA-treated leaves[b]	Seeds Leaves ABA-treated leaves[b]	No change seen
CDeT11–24	Not done	Seeds (embryos) Pollen No activity in leaves Not inducible by ABA except for very young seedlings	Seeds Activity inducible by ABA and dehydration in young leaves In root tips, inducible in roots	Increased responsiveness to ABA in leaves

[a]The references for the promoter analyses are Michel *et al.* 1993, 1994; Furini *et al.*, 1996; Velasco *et al.*, 1998.
[b]Abscisic acid (ABA)-treated leaves showed a higher activity than untreated leaves.

only transient homologous transformations were carried out and tobacco as well as *Arabidopsis* were used for stable transformations.

A common observation was that all three promoters were highly active in seeds and pollen. The studies revealed that apparently different DNA motifs were involved in gene activation by ABA. The CDeT6–19 promoter is more different from CDeT27–45 and CDeT11–24 as no *de novo* protein synthesis is required for ABA-mediated transcription, and promoter activities were inducible by dehydration or ABA in vegetative tissues of tobacco and *Arabidopsis*. Surprisingly, CDeT27–45 and CDeT11–24 were not active in vegetative tissues except for young seedlings and some root tissues in the case of CDeT11–24. The promoter activities of these two genes were thus mainly restricted to seeds and pollen, two types of tissues which undergo natural desiccation.

The ectopic expression of the *Arabidopsis ABI-3* gene product led to ABA inducibility in leaves of plants transformed with CDeT27–45 and CDeT11–24 promoter-*GUS* (*Table 2*); this suggested that possible regulatory factors are present in leaves of *C. plantagineum*, but missing in *Arabidopsis* or tobacco. One approach to isolate such factors was the cloning of the *vp1* or *ABI-3* gene homologue from *C. plantagineum* (Chandler and Bartels, 1997). Both *ABI-3* from *Arabidopsis* and *VP1* from maize are considered to be essential for completion of the seed maturation programme. Both gene products have been shown to transactivate LEA genes in transient transformation assays or in transgenic plants. The results displayed in *Table 2* supported the hypothesis that in vegetative tissues of *C. plantagineum* an ABI-3/vp1-like gene product may exist which is important for the activation of LEA genes.

The gene *CpVp1* was isolated from *C. plantagineum*. *CpVp1* had a very similar gene structure to homologues from other plants. However, its expression was restricted to seeds and it was not active in mature vegetative tissues, where many LEA genes are expressed. Therefore other factors must be involved in regulating desiccation-induced gene expression in leaves. It is an extremely important goal for the future to identify genes which act as master switches and control the activation of LEA genes. Such genes could pave the way to manipulate stress tolerance in higher plants for which cold tolerance in *Arabidopsis* is the first excellent example (Jaglo-Ottensen *et al.*, 1998).

References

Allamillo, J.M., Roncarati, R., Heino, P., Velasco, R., Nelson, D., Elster, R., Bernacchia, G., Furini, A., Schwall, G., Salamini, F. and Bartels, D. (1994) Molecular analysis of desiccation tolerance in barley embryos and in the resurrection plant *Craterostigma plantagineum*. *Agronomie* **2**: 161–167.

Artus, N.N., Uemura, M., Steponkus, P.L., Gilmour, S.J., Lin, C. and Thomashow, M.F. (1996) Constitutive expression of the cold-regulated *Arabidopsis thaliana COR15a* gene affects both chloroplast and protoplast freezing tolerance. *Proc. Natl Acad. Sci. USA* **93**: 13404–13409.

Asghar, R., Fenton, R.D., DeMason, D.A. and Close, T.J. (1994) Nuclear and cytoplasmic localization of maize embryo and aleurone dehydrin. *Protoplasma* **177**: 87–94.

Baker, J., Steele, C. and Dure, L. III. (1988) Sequence and characterization of 6 *Lea* proteins and their genes from cotton. *Plant Mol. Biol.* **11**: 277–291.

Bockel, C., Salamini, F. and Bartels, D. (1998) Isolation and characterization of genes expressed during early events of the dehydration process in the resurrection plant *Craterostigma plantagineum. J. Plant Physiol.* **152:** 158–166.

Busk, P.K., Jensen, A.B. and Pagès, M. (1997) Regulatory elements in vivo in the promoter of the abscisic acid responsive gene *rab 17* from maize. *Plant J.* **11:** 1285–1295.

Chandler, J.W. and Bartels, D. (1997) Structure and function of the *vp1* gene homologue from the resurrection plant *Craterostigma plantagineum* Hochst. *Mol. Gen. Genet.* **256:** 539–546.

Close, T.J. (1996) Dehydrins: emergence of a biochemical role of a family of plant dehydration proteins. *Physiol. Plant.* **97:** 795–803.

Close, T.J., Fenton, R.D. and Moonan, F. (1993) A view of plant dehydrins using antibodies specific to the carboxy terminal peptide. *Plant Mol. Biol.* **23:** 279–286.

Dure, L. III, Greenway, S.C. and Galau, G.A. (1981) Developmental biochemistry of cotton seed embryogenesis and germination. XIV. Changing mRNA populations as shown by in vitro and in vivo protein synthesis. *Biochem.* **20:** 4162–4168.

Dure, L. III, Crouch, M., Harada, J., Ho, T.-H.D., Mundy, J., Quatrano, R., Thomas, T. and Sung, Z.R. (1989) Common amino acid sequence domains among the LEA proteins of higher plants. *Plant Mol. Biol.* **12:** 475–486.

Furini, A., Parcy, F., Salamini, F. and Bartels, D. (1996) Differential regulation of two ABA-inducible genes from *Craterostigma plantagineum* in transgenic *Arabidopsis* plants. *Plant Mol. Biol.* **30:** 343–349.

Goday, A., Jensen, A.B., Culianez-Marcia, F.A., Albà, M.M., Figueras, M., Serratosa, J., Torrent, M. and Pagès, M. (1994). The maize abscisic acid-responsive protein Rab17 is located in the nucleus and interacts with nuclear localization signals. *Plant Cell* **6:** 351–360.

Guiltinan, M.J., Marcotte, W.R. and Quatrano, R.S. (1990) A plant leucine zipper protein that recognizes an abscisic acid response element. *Science* **250:** 267–271.

Imai, R., Chang, L., Ohta, A., Bray. E. and Takagi, M. (1996) A lea-class gene of tomato confers salt and freezing tolerance when expressed in *Saccharomyces cerevisiae. Gene* **170:** 243–248.

Iturriaga, G., Schneider, K., Salamini, F. and Bartels, D. (1992) Expression of desiccation-related proteins from the resurrection plant *Craterostigma plantagineum* in transgenic tobacco. *Plant Mol. Biol.* **20:** 555–558.

Jaglo-Ottensen, K.R., Gilmour, S.J., Zarka, D.G., Schabenberger, O. and Tomashow, M.F. (1998) Arabidopsis CBF1 overexpression induces *COR* genes and enhances freezing tolerance. *Science* **280:** 104–106.

Kim, S.Y., Chung, H.-J. and Thomas, T.L. (1997) Isolation of a novel class of bZIP transcription factors that interact with ABA-responsive and embryo-specification elements in the Dc3 promoter using a modified yeast one-hybrid system. *Plant J.* **11:** 1237–1251.

Leung, J. and Giraudat, J. (1998) Abscisic acid signal transduction. *Annu. Rev. Plant Physiol. Plant Mol. Biol.* **49:** 199–222.

Lisse, T., Bartels, D., Kalbitzer, H.R. and Jaenicke, R. (1996) The recombinant dehydrin-like desiccation stress protein from the resurrection plant *Craterostigma plantagineum* displays no defined three-dimensional structure in its native state. *Biol. Chem.* **377:** 555–561.

Liu, Q., Kasuga, M., Sakuma, Y., Abe, H., Miura, S., Yamaguchi-Shinozaki, K. and Shinozaki, K. (1998) Two transcription factors, DREB1 and DREB2, with an EREBP/AP2 DNA binding domain separate two cellular signal transduction pathways in drought- and low-temperature-responsive gene expression, respectively, in Arabidopsis. *Plant Cell* **10:** 1–17.

Michel, D., Salamini, F., Bartels, D., Dale, P., Baga, M. and Szalay, A. (1993) Analysis of a desiccation and ABA-responsive promoter isolated from the resurrection plant *Craterostigma plantagineum. Plant J.* **4:** 29–40.

Michel, D., Furini, A., Salamini, F. and Bartels, D. (1994) Structure and regulation of an ABA- and desiccation-responsive gene from the resurrection plant *Craterostigma plantagineum. Plant Mol. Biol.* **24:** 549–560.

Nakagawa, H., Ohmiya, K. and Hattari, T. (1996) A rice bZIP protein, designated OSBZ8, is rapidly induced by abscisic acid. *Plant J.* **9:** 217–227.

Nelson, D., Salamini, F. and Bartels, D. (1994) Abscisic acid promotes novel DNA-binding activity to a desiccation-related promoter of *Craterostigma plantagineum. Plant J.* **5:** 451–458.

Oeda, K., Salinas, J. and Chua, N.-H. (1991) A tobacco bZIP transcription activator (TAF-1) binds to a G-box-like motif conserved in plant genes. *EMBO J.* **10:** 1793–1802.

Piatkowski, D., Schneider, K., Salamini, F. and Bartels, D. (1990) Characterization of five abscisic acid-responsive cDNA clones isolated from the desiccation-tolerant plant *Craterostigma plantagineum* and their relationship to other water-stress genes. *Plant Physiol.* **94:** 1682–1688.

Roberts, J.K., DeSimone, N.A., Lingle, W.L. and Dure, L. III. (1993) Cellular concentrations and uniformity of cell-type accumulation of two Lea proteins in cotton embryos. *Plant Cell* **5:** 769–780.

Schneider, K., Wells, B., Schmelzer, E., Salamini, F. and Bartels, D. (1993) Desiccation leads to the rapid accumulation of both cytosolic and chloroplastic proteins in the resurrection plant *Craterostigma plantagineum* Hochst. *Planta* **189:** 120–131.

Shen, Q.X., Zhang, P.H. and Ho, T.-H.D. (1996) Modular nature of abscisic acid (ABA) response complexes: composite promoter units that are necessary and sufficient for ABA induction of gene expression in barley. *Plant Cell* **8:** 1107–1119.

Skriver, K., Olsen, F.L., Rogers, J.C. and Mundy, J. (1991) *Cis*-acting DNA elements responsive to gibberellin and its antagonist abscisic acid. *Proc. Natl Acad. Sci. USA* **88:** 7266–7270.

Steponkus, P.L., Uemura, M., Joseph, R.A., Gilmour, S.J. and Tomashow, M.F. (1998) Mode of action of the *COR15a* gene on the freezing tolerance of *Arabidopsis thaliana*. *Proc. Natl Acad. Sci. USA* **95:** 14570–14575.

Velasco, R., Salamini, F. and Bartels, D. (1998) Gene structure and expression analysis of the drought- and abscisic acid-responsive CDeT11–24 gene family from the resurrection plant *Craterostigma plantagineum* Hochst. *Planta* **204:** 459–471.

Xu, D., Duan, X., Wang, B., Hong, B., Ho, T.-H.D. and Wu, R. (1996) Expression of a late embryogenesis abundant protein gene, *HVA1*, from barley confers tolerance to water deficit and salt stress in transgenic rice. *Plant Physiol.* **110:** 249–257.

Yamaguchi-Shinozaki, K., Urao, T. and Shinozaki, K. (1995) Regulation of genes that are induced by drought stress in *Arabidopsis thaliana*. *J. Plant Res.* **108:** 127–136.

Chapter 20

Proteins correlated with desiccation tolerance in a resurrection grass *Sporobolus stapfianus*

Robert D. Gianello, Jianbo Kuang, Donald F. Gaff, Hamid R. Ghasempour, Cecilia K. Blomstedt, John D. Hamill and Alan D. Neale

1. Background

The monocotyledonous species *Sporobolus stapfianus* is notable for a desiccation tolerance similar to that of the resurrection dicotyledonous species *Craterostigma plantagineum* (Gaff, 1971; Gaff and Ellis, 1974). Changes in gene expression occur in *C. plantagineum* leaves as desiccation tolerance is induced in the plants during drying; novel transcripts arising during drying indicate six genes that are correlated with desiccation tolerance in *C. plantagineum* (Bartels *et al.*, 1990; Michel *et al.*, 1993; Piatkowski *et al.*, 1990). The resurrection grass *S. stapfianus* provides an opportunity for us to investigate whether changes in gene expression are similarly implicated in desiccation tolerance of a monocot and the extent to which such changes may be reflected in the protein complement of drying plants. Analysis of mRNA indicated 12 nuclear-encoded transcripts correlated with desiccation tolerance in *S. stapfianus* (Blomstedt *et al.*, 1998; Gaff *et al.*, 1993). Our study investigates the extent of changes in the complement of proteins synthesized in leaves on drying plants that are undergoing induction of desiccation tolerance and compares these changes with those in drying detached leaves of *S. stapfianus* (which remain desiccation sensitive) and with changes in protein complement in drying leaves of the desiccation-sensitive grass *S. pyramidalis*. Identification is attempted of the functional role of two drought-induced proteins that are correlated with induction of desiccation tolerance in *S. stapfianus*.

2. Results

2.1 In vivo proteins

Novel *in vivo* proteins begin to appear at about 85% relative water content (RWC),

Plant Responses to Environmental Stress, edited by M.F. Smallwood, C.M. Calvert and D.J. Bowles.

Table 1. Number of proteins (translated *in vivo*) first seen as novel, changed or lost at each relative water content (RWC) in leaves drying on intact plants of *S. stapfianus*

RWC (%)	Novel	(More)	(Less)	Lost	Total changes at RWC
95	0	0	0	0	0
85	4	0	0	2	6
73	2	1	0	1	4
60	3	1	1	0	5
51	1	0	8	0	9
37	13	6	4	2	25
3.5	2	2	0	2	6
Total	25	10	13	7	55

Plant material was grown and dried as in Blomstedt *et al.* (1998). Control samples of fully hydrated leaf tissue were taken simultaneously with samples of drought-stressed leaves. Protein complements in fully hydrated leaves, in attached-drying leaves and in detached-drying leaves were compared following extraction and partitioning by two-dimensional polyacrylamide gel electrophoresis (2D PAGE) (Bartels *et al.*, 1988).

Table 2. Changes in the protein complement during drying of leaves of *Sporobolus* spp

Type of change	S. stapfianus		S. pyramidalis
	Intact plants (desiccation tolerant)	Detached leaves (desiccation sensitive)	Intact plants (desiccation sensitive)
A. Novel protein	25	3	4
B. Increased abundance	10	4	1
C. Decreased abundance	13	1	8
D. Protein lost	7	4	19
Totals	55	12	32
Ratio (A+B)/(C+D)	1.8	1.4	0.2

In the case of the resurrection grass *Sporobolus stapfianus* Gandoger, leaves which dehydrate while they are attached to intact drying plants survive desiccation, whereas leaves which dry after being cut from well-hydrated plants are killed before they reach air-dryness (Gaff and Loveys, 1993). Fully hydrated detached leaves were allowed to dry to the appropriate relative water contents (RWCs), then were kept in air of 100% RWC to prevent further water loss from these samples until leaves on intact plants had dried to comparable RWC so that control and test samples were extracted at the same time. Leaves of the desiccation-sensitive grass *Sporobolus pyramidalis* Beauv. were sampled at appropriate water contents as plants dried intact under the same conditions as for *S. stapfianus* (*Table 1*).

during drying of intact plants of *S. stapfianus,* which survive desiccation (*Table 1*). This mild stress level lies well before the main rise in drought-induced abscisic acid (ABA) in *S. stapfianus* (a stress-induced hormone associated with stress resistance in other species) (Gaff and Loveys, 1993).

Detached leaves (which are injured by desiccation) show no novel proteins when the leaves are dried rapidly, and only three novel proteins when the leaves are dried slowly (*Table 2*). The three novel proteins differed from those induced in leaves that had dried attached.

In the desiccation-sensitive grass *S. pyramidalis*, only four novel proteins appeared during drying, but a large number (*ca* 26) of proteins were lost or reduced in abundance (cf. 20 in intact dried *S. stapfianus*) (*Table 2*).

Two main phases of change in *in vivo*-synthesized proteins can be distinguished in 'viable' drying of intact *S. stapfianus* plants: 10 novel proteins appeared in the 85–51% RWC range; 13 novel proteins arose in the second phase (37–3% RWC, presumably mainly at the higher section of this range) (*Table 1*). A brief intermediate phase occurs around 50% RWC with a decrease in the abundance of eight proteins.

Since many novel specific mRNA species arise in *S. stapfianus* leaves (attached) at moderate and severe drought stress (Gaff *et al.*, 1993), the novel proteins that are synthesized *in vivo* probably result in many cases from stress-induced activation of the genes encoding them.

2.2 Sequence data

Two novel proteins, which arose at 83% RWC and persisted throughout drying were purified for sequencing of amino acid residues. Two-dimensional polyacrylamide gel electrophoresis (2D PAGE) showed that the two major proteins were Novel Protein nos 1 and 2 (SALD1 and SALD2; Kuang *et al.*, 1995). SALD1 peptides 8 and 9 generated the following sequences, respectively:

SALDl Peptide 8:
Tyr-Ala-Ala-Ala-Val-Pro-Thr-Glu-Tyr-Gly-Lys

SALD1 Peptide 9:
Leu-Gly-Asp-Leu-Gly-Gly-Ser-Ala-Val-Glu-Asp-Pro-Ala-Ala-Pro

SALD2 Peptide 7 produced the sequence:
Tyr-Ala-Ala-Ala-Val-Pro-Thr-Gln-Tyr-Gly-Arg

2.3 Identification of sequences

The three sequences were compared to known protein sequences using the computer databases GenePep, GenePepWeekly, SwissPro and SwissWeekly. SALDl Pep 8 had 100% identity over a six consecutive amino acid sequence with galactose-1-phosphate uridylyltransferase reductase (*Figure 1*). SALD2 Pep 7 had 100% identity over a six consecutive amino acid stretch (last six of the above series) with DNA-dependent RNA polymerase (EC 2.7.7.6). Work is continuing to verify and strengthen the identification, given that the matches, although 100%, were over only six amino acids.

The sequence shared between these two proteins (SALD1 and SALD2), that is Tyr-Ala-Ala-Ala-Val-Pro-Thr, and their similar migration in 2D PAGE suggest that they are in the same family of proteins. A search of the protein sequence databases for the shared

```
                         1 0
SALD1:peptide 8      YAAAVPTEYGK
                     •|| || | |•
GAL7_BACSU       AKNVHWTVPTEYGELEM
                   1 4 0        15 0
```

Figure 1 Similarity of SALD1: peptide 8 to GAL7_BACSU = galactose-1-phosphate uridylyltransferase (EC 2.7.7.10) 513 AA.

sequence revealed best similarity with α-amylase and isoamylase and lesser similarity with RNA polymerase and glucoamylase. Possibly this protein segment binds to sugar moieties, including perhaps the ribose of RNA.

Proteome Analysis, in co-operation with the Macquarie University, is being applied to freshly purified samples of the two proteins (full-length SALD1 and SALD2) as an additional indicator of the identity of the two proteins. A preliminary analysis of SALD1 indicated highest similarity with amylase, but antibodies to α-amylase and to β-amylase hybridized only with a protein of much lower molecular size than SALD1 in western analyses.

When a cDNA library derived from *S. stapfianus* leaves was screened for clones encoding transcripts whose abundance increased during drying of the foliage, 55 clones encoded protein with similarity to dehydrin and four clones encoded protein with similarity to LEA (late embrogenesis abundant) group 3; three other similar-clone groupings contained two to three clones each; all other clones differed from each other (Blomstedt *et al.*, 1998). Northern analyses indicated that drought produced large increases in transcripts encoding dehydrin and LEA gp3 in leaves drying on intact plants (desiccation tolerant), but much smaller increases in desiccation-sensitive tissue (*S. stapfianus* leaves drying detached and *S. pyramidalis* leaves) (*Table 3*).

3. Discussion

3.1 Intact S. stapfianus

Drought resulted in numerous changes in the protein complement of leaves: 35 proteins were novel or of increased abundance during drying and 20 were lost or reduced in abundance. Protein changes then reflect the extensive changes in transcript, although the

Table 3. Levels of specific mRNA transcripts in drying leaves of *Sporobolus* spp.

Plant material (and desiccation tolerance status)	Probe for[*]	Northern signal at the % RWC indicated (% of maximum signal of all samples for the probe)	
		Fully hydrated	Maximum signal, at RWC
S. stapfianus			
Dried intact	Dehydrin	*ca* 1%	100%, at 10% RWC
(des. tolerant)	LEA3	*ca* 1%	100%, at 20–39% RWC
Dried detached	Dehydrin	*ca* 1%	*ca* 17%, at 20–39% RWC
(des. sensitive)	LEA3	< 1%	*ca* 10%, at 20–39% RWC
S. pyramidalis			
Dried intact	Dehydrin	*ca* 1%	*ca* 30%, at 20–39% RWC
(des. sensitive)	LEA3	< 1%	*ca* 38%, at <10% RWC

[*]Dehydrin, SDG1-8a; LEA3, SDG2-24 from *S. stapfianus* (Blomstedt *et al.*, 1998).
RNA was extracted from leaves of fully hydrated and of drying leaves, applied to filters and probed in northern analyses that were quantified using phospho-imaging (all as described in Blomstedt *et al.*, 1998).

novel or more abundant transcripts were more numerous (54 transcripts; Gaff *et al.*, 1993) than novel – plus increased – proteins.

Four novel proteins were detected even at slight drought stress (85% RWC). Two phases of protein changes were discernible: 10 novel proteins arose in the 85–51% RWC range and 15 other novel proteins in the 37–3% RWC range. The first phase includes the stress level 60% RWC at which attached leaves attain the ability to survive when subsequently they are detached then air-dried (Gaff and Loveys, 1993). Some of the protein changes in phase 1 therefore may be important in the later survival of low RWC. The phase 2 range of drought stress coincides with a marked accumulation of ABA in leaves on intact *S. stapfianus* plants (Gaff and Loveys, 1993). ABA can promote expression of a number of drought-inducible genes in *C. plantagineum* and in *S. stapfianus* (Bartels *et al.*, 1990; Blomstedt *et al.*, 1998). It is not surprising then that a number of genes which are activated genes in the 37–3% RWC phase are responsive to ABA (Blomstedt *et al.*, 1998). Metabolism in this very severe stage of drought stress must prevent injury that occurs in most angiosperm plants in the region of 30% RWC, and also ensure that mechanisms are in place to support the recovery of metabolism during subsequent rehydration, cf. the rehydrin proteins and mRNA that persist in the desiccated state of the moss *Tortula ruralis* to contribute to recovery during rehydration (Oliver and Bewley, 1984).

The identity of the two major novel proteins at 37% RWC remains unclear at this point. The putative enzymatic functions could contribute to tissue survival by conversion of starch to protective sugars or contribute to protein synthesis. Additional research is required to clarify the roles of the two proteins.

3.2 Desiccation sensitive leaves

The low number of drought-induced protein changes in leaves susceptible to drought injury (detached *S. stapfianus* leaves and intact *S. pyramidalis*) strengthens the view that the novel proteins and proteins of increased abundance in *S. stapfianus* plants drying intact are implicated in the induction of desiccation tolerance rather than reflecting injury-related processes (*Table 2*). The sensitivity of *S. pyramidalis* to desiccation may be related to the high proportion of proteins that decrease in abundance during drying compared to those that increase (*Table 2*).

Dehydrins and LEA3 transcripts are synthesized in both desiccation-tolerant and desiccation-sensitive tissue of *Sporobolus*, but dehydrin and LEA3 transcript abundances correlate well with desiccation tolerance (*Table 3*). This correlation strengthens the view that abundant accumulation of the encoded proteins assists the induction of desiccation tolerance during drying of *S. stapfianus*.

4. Conclusion

The above data together with earlier studies on mRNA in *S. stapfianus* support the hypothesis that a range of genes are activated during drying of plants of this resurrection grass and that the encoded proteins are implicated in the induction of desiccation tolerance. They further indicate that there are two main phases of alteration in gene expression, both of which are important to desiccation tolerance. Accumulation of ABA is probably important in the induction of the many novel proteins synthesized in the latter phase. Dehydrin and LEA3 transcripts in that phase correlate well with desiccation tolerance.

Acknowledgements

The authors thank the Meat Research Corporation (UMON.004) and the ARC (A19230441) for the financial support that made the project possible.

References

Bartels, D., Singh, M. and Salamini, F. (1988) Onset of desiccation tolerance during development of the barley embryo. *Planta* **175:** 485–492.
Bartels, D., Schneider, K., Terstappen, G., Piatkowski, D. and Salamini, F. (1990) Molecular cloning of abscisic acid-modulated genes which are induced during desiccation of the resurrection plant *Craterostigma plantagineum. Planta* **181:** 27–34.
Blomstedt, C.K., Gianello, R.D., Hamill, J.D., Neale, A.D., and Gaff, D.F. (1998) Drought-stimulated genes correlated with desiccation tolerance of the resurrection grass *Sporobolus stapfianus. Plant Growth Regul.* **24:** 153–161.
Gaff, D.F. (1971) The desiccation tolerant higher plants of southern Africa. *Science* **174:** 1033–1034.
Gaff, D.F. and Ellis, R.P. (1974) Southern African grasses with foliage that revives after dehydration. *Bothalia* **11:** 305–308.
Gaff, D.F. and Loveys, B.R. (1993) Abscisic acid levels in drying plants of a resurrection grass. *Trans. Malaysian Soc. Plant Physiol.* **3:** 286–287.
Gaff, D.F., Bartels, D. and Gaff, J.L. (1993) Changes in gene expression during drying in a desiccation tolerant grass *Sporobolus stapfianus* and a desiccation sensitive grass *Sporobolus pyramidalis. Aust. J. Plant Physiol.* **24:** 617–612.
Kuang, J., Gaff, D.F., Gianello, R., Blomstedt, C., Neale, A.D. and Hamill, J.D. (1995) Changes in in vivo protein complements in drying leaves of the desiccation-tolerant grass *Sporobolus stapfianus* and the desiccation-sensitive grass *Sporobolus pyramidalis. Aust. J. Plant Physiol.* **22:** 1027–1034.
Michel, D., Salamini, F., Bartels, D., Dale, P., Baga, M. and Szalay, A. (1993) Analysis of a desiccation and ABA responsive promoter isolated from the resurrection plant *Craterostigma plantagineum. Plant J.* **4:** 29–40.
Oliver, M.J. and Bewley, J.D. (1984) Plant desiccation and protein synthesis. V. Stability of poly(a)− and poly(a)+RNA during desiccation and their synthesis upon rehydration in the desiccation-tolerant moss *Tortula ruralis* and the intolerant moss *Craterineuron filicinum. Plant Physiol.* **74:** 917–922.
Piatkowski, D., Schneider, K., Salamini, F. and Bartels, D. (1990) Characterization of five abscisic acid-responsive cDNA clones isolated from the desiccation tolerant plant *Craterostigma plantagineum* and their relationship to other water-stress genes. *Plant Physiol.* **94:** 1682–1688.

Chapter 21

Seed peroxiredoxins

Reidunn B. Aalen, Robin A. P. Stacy and Camilla Haslekås

1. Background

Both from an agronomical and a biological point of view the seed is a very significant and interesting object to study. Seeds are very important as food and feed. Seed viability, germination efficiency and storage ability are properties that have a vast impact on crop production. These crucial features of seeds are based on two remarkable phenomena – that most seeds survive desiccation and that they can go into a period of dormancy, a state characterized by the inability to germinate immediately in otherwise supportive conditions (Bewley, 1997).

The majority of angiosperm seeds suffer a dramatical reduction in tissue water content during the final stages of seed development. The mature seed will then enter a period of metabolic quiescence, which can last for weeks, months or even years. The dry seed must absorb water and resume respiration in order to germinate. Dormant seeds can also take up water and resume respiration although no growth will take place (Bewley and Black, 1994).

Both desiccation and respiration give rise to potentially damaging reactive oxygen species. Such species can have deleterious effects on membranes, proteins and DNA. Obviously, mechanisms are needed to protect the components of cells, which are to survive desiccation and later give rise to the new plant (Bewley and Black, 1994). A central issue in understanding the molecular biology of seed desiccation and dormancy is to identify these mechanisms.

Late embryogenesis abundant (LEA) proteins are speculated to play a role in desiccation protection (Baker *et al.*, 1988), and circumstantial evidence indicates that LEA proteins are critical also for the induction and/or maintenance of dormancy (Galau *et al.*, 1991). Some *Lea* genes are expressed exclusively in the embryo, while others are also expressed in the only other tissue surviving desiccation, the aleurone layer of the endosperm.

2. Regulatory and functional studies of desiccation- and dormancy-related seed proteins

Mutants have been identified showing relatively severe phenotypes with altered cotyledon morphology and/or no or little expression of genes normally active during the

Plant Responses to Environmental Stress, edited by M.F. Smallwood, C.M. Calvert and D.J. Bowles.
© 1999 BIOS Scientific Publishers, Oxford.

maturation and desiccation stages of seed development. These mutants are viviparous (i.e. able to germinate prior to harvesting) and desiccation intolerant. To understand the specific roles of LEA proteins during desiccation and dormancy, knowledge of the functions of these proteins are needed. The first *Lea* genes characterized, belonging to Groups 1, 2 and 3, were found to encode hydrophilic proteins (Close, 1996; Dure, 1993a, 1993b). More recently new groups of *Lea* families have been recognized, some of which have been described as 'atypical' since the encoded proteins are not hydrophilic. In some cases, cellular localization has been examined, functional analyses performed and phenotypic changes in transgenic plants or yeast over-expressing LEA proteins have been consistent with a protective role against desiccation-induced stresses (e.g. see review by Close, 1996). So far, there are no direct candidate genes for involvement in the maintenance or termination of dormancy, although a number of *Lea* genes have been shown to have a dormancy-related expression pattern (Li and Foley, 1997).

Identification and functional characterization desiccation- and dormancy-related seed proteins can be expected to have a large impact on future selection for or modulation of important seed properties. The particular contribution from our group to this field of study is the characterization of a seed-specific antioxidant called peroxiredoxin.

3. Seed peroxiredoxins are atypical LEA proteins

Peroxiredoxins represent a family of thiol-requiring antioxidants first characterized in yeast (Kim *et al.*, 1988), and later identified in organisms ranging from bacteria to humans (Chae *et al.*, 1994). The family can be divided into two subgroups, one with two conserved cysteine residues (2-Cys) and another with one conserved cysteine (1-Cys). We have cloned genes for the 1-Cys peroxiredoxins in barley (*Per1*) and *Arabidopsis thaliana* (*AtPer1*) (Haslekås *et al.*, 1998; Stacy *et al.*, 1996). The barley and the *Arabidopsis* peroxiredoxin amino acid sequences show 74% identity to each other, and share about 50% identity with the human homologue. In comparison, the *Arabidopsis* 2-Cys protein, which is found in green tissue (Baier and Dietz, 1997), shows an identity of only 30%.

We have confirmed antioxidant activity *in vitro* of the barley peroxiredoxin (Stacy *et al.*, 1996). A PER1 fusion protein was purified using an *Escherichia coli* expression system. Applying a mix-function oxidation system containing thiol, where DNA will be degraded due to radical attack, we showed that purified PER1 fusion protein will protect the DNA against degradation in a dose-dependent manner.

Per1, initially called B15C, was first identified as a <u>B</u>arley <u>al</u>eurone and <u>em</u>bryo (*Balem*) expressed transcript after differential screening of an aleurone cDNA library (Aalen *et al.*, 1994; Jakobsen *et al.*, 1989). The protein is not particularly hydrophobic, but the transcript has a *Lea*-like accumulation pattern during maturation and desiccation stages of seed development, and disappears rapidly upon germination. Lately, we have shown a similar expression pattern for the PER1 protein (Stacy *et al.*, 1999). Like other *Lea* transcripts *Per1* can be induced in immature embryos by abscisic acid (ABA) and osmotic agents. Based on these findings PER1 can be characterized as an atypical LEA protein. By northern and *in situ* hybridization we have shown that the *Arabidopsis* homologue *AtPer1* has a temporal and spatial expression pattern very similar to *Per1* (Haslekås *et al.*, 1998). This is one of the few transcripts so far shown to be expressed in the aleurone layer in *Arabidopsis*. Contrary to some *Lea* genes, *AtPer1* cannot be induced in vegetative tissues by ABA or drought (Haslekås *et al.*, 1998).

In many respects 1-Cys peroxiredoxins differ from other antioxidants: most enzymatic antioxidants are encoded by gene families, and are expressed in many tissues and developmental stages of the plant life cycle. In contrast, *Per1* and *AtPer1* are single-copy genes specifically expressed during seed development with tissue-specific expression patterns restricted to the only parts of the seed surviving desiccation (Haslekås *et al.*, 1998; Stacy *et al.*, 1996).

Since protection against free radicals is an important aspect of desiccation tolerance (Leprince *et al.*, 1993), we have suggested such a protective role for the seed-specific peroxiredoxins.

4. Seed peroxiredoxin genes are expressed in a dormancy-related manner

Another interesting aspect is the expression patterns of peroxiredoxin genes in imbibed dormant seeds. Aalen *et al.* (1994) identified *Per1* (then called B15C) as almost identical to the transcript pBS128 from *Bromus secalinus* which was clearly up-regulated during imbibition of dormant embryos (Goldmark *et al.*, 1992). Our studies in barley have shown that *Per1*-expression patterns in mature barley embryos during imbibition differ between cultivars with dissimilar levels of dormancy (Stacy *et al.*, 1996). In all cultivars, *Per1* is up-regulated during early imbibition. However, transcript levels increase during imbibition of dormant embryos, but decline rapidly upon germination of non-dormant embryos. Similar results were found in *Arabidopsis*. In freshly harvested seeds, which will not germinate, the transcript level is maintained during imbibition, but decreases in after-ripened germinating seeds (Haslekås *et al.*, 1996). Protein studies show that the PER1 protein disappears in germinating seeds, although slower than the transcript (Stacy *et al.*, 1999).

We have indications that gibberellic acid (GA) is involved in the suppression of *Per1* transcripts in germinating seeds, since the *Per1* transcript decreases rapidly in barley aleurone layers when mature embryoless seeds are incubated with GA_3 (Stacy *et al.*, 1999). Few transcripts have to date been shown to be GA-suppressible, while a number of genes important for the mobilization of storage compounds of the starchy endosperm during germination, most notably α-amylase genes, are known to be GA-inducible.

We have previously hypothesized that peroxiredoxins could be involved in the maintenance of dormancy in dormant seeds (Stacy *et al.*, 1996). Therefore, we investigated the expression level of *AtPer1* in *Arabidopsis* ABA-deficient (*aba-1*) and ABA-insensitive (*abi3-1*) mutants with reduced dormancy. In the *aba-1* mutant the *AtPer1* transcript level is similar to wild type, while in *abi3-1* the level is about 10% of wild type (Haslekås *et al.*, 1998). This result shows that a high *AtPer1* level is not sufficient to express dormancy, and that *AtPer1* expression during seed development seems to be controlled by an endogenous ABA-independent stage-specific pathway.

We have also suggested that the seed peroxiredoxin has a protective role during early imbibition and germination. Other antioxidants (e.g. superoxide dismutases and catalases) are first induced to high levels in the seed during germination, where they are assumed to protect against activated oxygen by-products of respiration. Respiration is, however, resumed already at the onset of imbibition and in hydrated dormant seeds. The expression patterns of *Per1* and *AtPer1* are consistent with the hypothesis that peroxiredoxins protect against oxidative damage in these situations, when the living seed tissues are not able to gain protection from germination-specific antioxidants.

5. Further investigations on seed peroxiredoxins

Our study of seed peroxiredoxins started in barley. In order to expand our studies we cloned the homologue in *Arabidopsis*, which offers several advantages over barley. A number of interesting mutant lines are available for investigation. *Arabidopsis* is an easily transformable model plant, and exhibits dormancy. Presently, transgenic *Arabidopsis* plants which (i) over-express PER1, (ii) express the antisensed *AtPer1* cDNA, and (iii) contain *AtPer1* promoter fragments coupled to the *gus* reporter gene, are under investigation. These transgenic plants will hopefully give information on functional and regulatory aspects of peroxiredoxin genes and proteins.

Lately we have shown, using antiserum against the PER1 protein, that PER1 is present in the nucleus with a high concentration over the nucleolus in developing embryo and aleurone cells (Stacy *et al.*, 1999). This opens the exciting possibility that seed peroxiredoxin antioxidants are protecting nuclear components of desiccation-tolerant cells.

References

Aalen, R.B., Opsahl-Ferstad, H.-G., Linnestad, C. and Olsen, O.-A. (1994) The transcripts encoding an oleosin and a dormancy-related protein are present in both the aleurone layer and the embryo of developing barley (*Hordeum vulgare* L.) seeds. *Plant J.* **5**: 385–396.

Baier, M. and Dietz, K.-J. (1997) The plant 2-Cys peroxiredoxin BAS1 is a nuclear-encoded chloroplast protein: its expressional regulation, phylogenetic origin, and implications for its specific physiological function in plants. *Plant J.* **12**: 179–190.

Baker, J., Steele, C. and Dure, L. III. (1988) Sequence and characterization of 6 *Lea* proteins and their genes from cotton. *Plant Mol. Biol.* **11**: 277–291.

Bewley, J.D. (1997). Seed germination and dormancy. *Plant Cell* **9**: 1055–1066.

Bewley, J.D. and Black, M. (1994) *Seeds: Physiology of Development and Germination.* Plenum Press, New York.

Chae, H.Z., Robison, K., Poole, L.B., Church, G., Storz, G. and Rhee, S.G. (1994) Cloning and sequencing of thiol-specific antioxidant from mammalian brain: alkyl hydroperoxide reductase and thiol-specific antioxidant define a large family of antioxidant enzymes. *Proc. Natl Acad. Sci. USA* **91**: 7017–7021.

Close, T.J. (1996) Emergence of a biochemical role of a family of plant dehydration proteins. *Physiol. Plant.* **97**: 795–803.

Dure, L. III. (1993a) Structural motifs in Lea proteins. In: Plant Responses to Cellular Dehydration during Environmental Stress: Current Topics in Plant Physiology, Vol. 10 (eds. T.J. Close and E.A. Bray). American Society of Plant Physiologists, MA, pp. 91–103.

Dure, L. III. (1993b) A repeating 11-mer amino acid motif and plant desiccation. *Plant J.* **3**: 363–369.

Seed peroxiredoxins

Galau, G.A., Jakobsen, K.S. and Hughes, D.W. (1991) The controls of late embryogenesis and early germination. *Plant Physiol.* **81:** 280–288.

Goldmark, P.J., Curry, J., Morris, C.F. and Walker-Simmons, M.K. (1992) Cloning and expression of an embryo-specific mRNA up-regulated in hydrated dormant seeds. *Plant Mol. Biol.* **19:** 433–441.

Haslekås, C., Stacy, R.A.P., Nygaard, V., Culiáñez-Macià, F.A. and Aalen, R.B. (1998) The expression of a peroxiredoxin antioxidant gene, *AtPer1*, in *Arabidopsis thaliana* is seed-specific and related to dormancy. *Plant Mol. Biol.* **36:** 833–845.

Jakobsen, K., Klemsdal, S.S., Aalen, R.B., Bosnes, M., Alexander, D. and Olsen, O.-A. (1989) Barley aleurone cell development: molecular cloning of aleurone-specific cDNAs from immature grains. *Plant Mol. Biol.* **12:** 285–293.

Kim, K., Kim, I.H., Lee, K.-Y., Rhee, S.G. and Stadtman, E.R. (1988) The isolation and purification of a specific 'protector' protein which inhibits enzyme inactivation by a thiol/Fe(III)/O$_2$ mixed-function oxidation system. *J. Biol. Chem.* **263:** 4704–4711.

Leprince, O., Hendry, G.A.F. and McKersie, B.D. (1993) The mechanisms of desiccation tolerance in developing seeds. *Seed Sci. Res.* **3:** 231–246.

Li, B. and Foley, M.E. (1997) Genetic and molecular control of seed dormancy. *Trends Plant Sci.* **2:** 384–389.

Stacy, R.A.P., Munthe, E., Steinum, T., Sharma, B. and Aalen, R.B. (1996) A peroxiredoxin antioxidant is encoded by a dormancy-related gene, *Per1*, expressed during late development in the aleurone and embryo of barley grains. *Plant Mol. Biol.* **31:** 1205–1216.

Stacy, R.A.P., Nordeng, T.W., Culiáñez-Macià, F.A. and Aalen, R.B. (1999) The dormancy-related peroxiredoxin antioxidant, PER1, is localized to the nucleus of barley embryo and aleurone cells. *Plant J.* (in press).

Chapter 22

PAP phosphatase, an enzyme conserved throughout evolution: its role in lithium and sodium toxicities

Pedro L. Rodríguez, José M. López-Coronado, Rosario Gil-Mascarell, José R. Murguía and Ramón Serrano

1. Background: Hal2, a lithium- and sodium-sensitive PAP phosphatase from *Saccharomyces cerevisiae*

HAL2 was the first gene to be biochemically characterized as a PAP phosphatase (Murguia *et al.*, 1995), that is a magnesium-dependent enzyme that specifically hydrolyses the 3'-phosphate from 3'-phosphoadenosine 5'-phosphate (PAP), thereby recycling PAP to adenosine 5'-monophosphate (AMP) (see *Figure 1*). *HAL2* was isolated in a screening for *Saccharomyces cerevisiae* genes which upon increase in gene dosage improve growth under salt stress (Glaser *et al.*, 1993). Analysis of the sequence of *HAL2* identified two sequence motifs which are also conserved in the family of inositol monophosphatases, however, the sequence of Hal2 does not share an overall similarity to inositol monophosphatases. This fact suggested that Hal2 could be a phosphatase and posed the question of the actual substrate of the enzyme. A second clue for the elucidation of the function of *HAL2* came from the analysis of the *HAL2* gene disruption. The only apparent phenotype of *HAL2*-disrupted cells is an auxotrophy for methionine (Glaser *et al.*, 1993). Therefore, *HAL2* must play a crucial role in methionine biosynthesis. Complementation analysis with mutants defective in different genes of the methionine pathway identified *HAL2* as identical to MET22 (Thomas *et al.*, 1992). Curiously, Hal2 does not correspond to any of the metabolic enzymes involved in methionine biosynthesis. Instead, it plays a crucial role in a side reaction (see below) which is essential for this metabolic pathway.

Methionine biosynthesis in yeast requires a reduction of inorganic sulphate prior to assimilation into carbon compounds (Jones and Fink, 1982). This process is initiated by the reaction of sulphate with ATP (catalysed by ATP sulphurylase) to produce adenosine 5'-phosphosulphate (APS), which is phosphorylated to 3'-phosphoadenosine 5'-phosphosulphate (PAPS) by APS kinase. Therefore, PAPS represents the activated form

Plant Responses to Environmental Stress, edited by M.F. Smallwood, C.M. Calvert and D.J. Bowles.

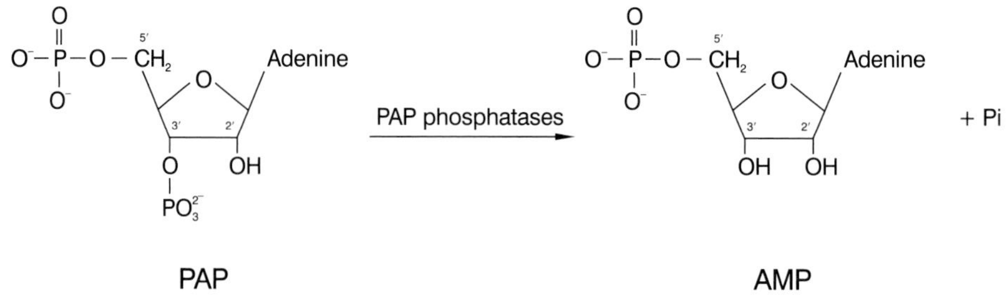

Figure 1. Diagram of the reaction catalysed by 3'-phosphoadenosine 5'-phosphate (PAP) phosphatase. PAP phosphatases prevent the accumulation of PAP, a potent inhibitor of sulphotransferases and RNA-processing enzymes. PAP phosphatases hydrolyse the 3'-phosphate from PAP, thereby recycling PAP to adenosine 5'-monophosphate (AMP). 3'-Phosphoadenosine 5'-phosphosulphate (PAPS), 2'-PAP, and to some extent inositol 1,4-bisphosphate, are also accepted by the enzyme, although with different preference depending on the PAP phosphatase considered (Gil-Mascarell *et al.*, 1999; Lopez-Coronado *et al.*, 1999; Quintero *et al.*, 1996).

of sulphate which will be reduced to sulphite by PAPS reductase, generating PAP as a side product. PAP exerts a product inhibition on PAPS reductase. Consequently, a fast removal of PAP is necessary to maintain active PAP reductase and hence, methionine biosynthesis. A specific phosphatase could perform such a role, for example removal of the 3'-phosphate from PAP would recycle PAP to AMP. This clue prompted us to examine whether the Hal2 phosphatase might be such an enzyme. Effectively, the biochemical analysis of Hal2 demonstrated that PAP served as a substrate of Hal2. HPLC analysis of this reaction confirmed the specific conversion of PAP to AMP by hydrolysis of the 3'-phosphate (Murguia *et al.*, 1995), with a K_m value of 3 µM.

Why does over-expression of *HAL2* suppress salt toxicity? Biochemical characterization of Hal2 shows that it is a lithium- and sodium-sensitive enzyme (*Table 1*). As this salt-sensitive enzyme catalyses an important metabolic reaction, it constitutes a primary target of sodium toxicity in yeast. Exposure of yeast cells to toxic concentrations of sodium produces an inhibition of the Hal2 enzyme and consequently PAP accumulates

Table 1. Sensitivity to lithium and sodium of 3'-phosphoadenosine 5'-phosphate (PAP) phosphatases from different organisms

Enzyme	IC_{50} Li$^+$ (mM)	IC_{50} Na$^+$ (mM)
ScHal2	0.1	20
AtAHL	10	45
AtSAL1	0.1	200
AtSAL2	10	200
RnPIP	0.8	Not sensitive

The concentration either of LiCl or NaCl that decreased the activity of the enzyme by 50% relative to a reaction without salt (IC_{50}) was estimated. The PAP phosphatase assay was performed with 0.2 mM PAP, under conditions of maximal activity, pH 7.5 and 1 mM Mg^{2+}, for the *Saccharomyces cerevisiae* Hal2 (ScHal2), *Arabidopsis thaliana* (AtAHL, AtSAL1, and AtSAL2) and *Rattus norvegicus* (RnPIP) enzymes.

(Murguia *et al.*, 1996). As a domino effect, PAPS reductase becomes inhibited and methionine biosynthesis is compromised. Accordingly, *hal2* cells are auxotrophic for methionine, and the lithium and sodium tolerance of yeast cells is improved by methionine supplementation. Over-expression of *HAL2* decreases the lithium and sodium toxicity by the intracellular increase of PAP phosphatase activity.

Yet an unsolved question remained in this scenario. If the sole effect of PAP accumulation was inhibition of PAPS reductase, supplementation with sulphite (the product of PAPS reductase) should bypass the requirement for *HAL2*. In contrast, *hal2* mutants fail to grow with sulphite as the sulphur source. We speculated that PAP accumulation might have a second toxic effect on a target unidentified at that moment. Confirmation for this proposal has come from the recent identification of PAP-sensitive RNA-processing RNAases (5′→3′ exoribonucleases) by Dichtl *et al.* (1997). Since PAP is a nucleoside 3′,5′-bisphosphate, it mimics the monomers of a nucleic acid chain and accumulation of PAP might prevent the attack to the phosphodiester bond of the PAP-sensitive 5′→3′ exoribonucleases. Therefore, the PAP phosphatase activity of Hal2 is required not only for methionine biosynthesis but also for RNA processing. Methionine supplementation of *hal2* cells, in addition to restoring the supply of organic sulphur, prevents the accumulation of PAP by repression of the genes involved in sulphate assimilation. Instead, the feedback repression due to added sulphite is not as strong as with methionine, PAP accumulates and as a result, RNA processing is inhibited.

2. The occurrence of PAP phosphatases in other organisms

PAPS is a universal donor of activated sulphate. PAP is generated either as a side product of PAPS reductase enzymes or during the biosynthesis of sulpho-conjugated molecules. Product inhibition by PAP appears to be a general phenomenon in enzymes that utilize PAPS as co-substrate, as exemplified by PAPS reductases from *Escherichia coli* and *S. cerevisiae* (Schwenn and Schriek, 1987; Schwenn *et al.*, 1988), plant flavonol sulphotransferases (Varin and Ibrahim, 1992) and desulphoglucosinolate sulphotransferases (Glendening and Poulton, 1990; Jain *et al.*, 1990), or animal sulphotransferases (Roth *et al.*, 1982). Therefore, PAP phosphatases must have been conserved throughout evolution and probably they act co-ordinately with enzymes that utilize PAPS. The PAP phosphatases, as exemplified by Hal2, are likely cation-sensitive phosphatases. However, the degree of sensitivity to monovalent cations will probably differ depending on the organism. For instance, animal cells are adapted to living with a high extracellular sodium concentration. In contrast, high sodium concentrations are toxic for glycophytic plants. This fact could be reflected in a different sensitivity to sodium between plant and animal PAP phosphatases. Obviously, the identification of lithium- and sodium-sensitive enzymes is of interest in medicine and agriculture, as lithium-sensitive enzymes are the targets of lithium therapy in humans and plant sodium-sensitive enzymes might constitute potential targets of salt toxicity in crops. Therefore, we have explored the presence of this family of enzymes both in plants and animals. For this purpose, we have used mainly functional complementation of the yeast *hal2* mutant by cDNAs either from *Arabidopsis thaliana* or *Rattus norvegicus*.

3. Identification of new plant PAP phosphatases

AtSAL1 was the first plant PAP phosphatase to be identified (Quintero *et al.*, 1996). This gene was cloned by expression in yeast of a cDNA library from *Arabidopsis* and selection of those clones that confer Li$^+$ tolerance. We have recently characterized two new PAP phosphatases from *Arabidopsis*, *AtSAL2* and *AtAHL*, which complement the auxotrophy for methionine of *hal2* cells (Gil-Mascarell *et al.*, 1999). SAL2 shows 68% amino acid sequence identity to the previously identified SAL1. AHL is only 50% identical to SAL1 and 46% identical to SAL2. The AHL and SAL2 recombinant proteins catalyse the conversion of PAP to AMP in a magnesium-dependent reaction sensitive to inhibition by Ca^{2+} and Li$^+$ ions. The PAP phosphatase activity of AHL is sensitive to physiological concentrations of Na$^+$, whereas the activities of SAL1 and SAL2 are not (*Table 1*). Therefore, AHL is sodium-sensitive and constitutes a potential target of salt toxicity.

   As discussed above, the PAP phosphatase activity of Hal2 is crucial for methionine biosynthesis. In plants, recent results (Gutierrez-Marcos *et al.*, 1996; Setya *et al.*, 1996) indicate that reductive sulphate assimilation proceeds via the reduction of APS to sulphite by APS reductase and the subsequent reduction of sulphite to sulphide by sulphite reductase. In other words, plants do not seem to have PAPS reductases generating PAP and consequently, PAPS is not a direct intermediate in the pathway. Instead PAPS is required as a source of activated sulphate in sulphation reactions. Sulphate conjugation in plants is emerging as an important reaction in very different processes (Varin *et al.*, 1997a). Sulphation is critical to the function of a wide variety of biomolecules. For instance, a sulphate moiety is required for the biological activity of gallic acid glucoside sulphate in the seismonastic response of *Mimosa pudica* (Varin *et al.*, 1997b). Sulphated flavonoids may play a role in the regulation of plant growth by interfering with the auxin receptor (Faulkner and Rubery, 1992). Choline sulphate accumulates in response to salinity or drought stress (Hanson *et al.*, 1994). Glucosinolates are sulphated compounds which contribute to the plant defence against pathogen infection (Mithen, 1992). Thus, the sulphation of such a diverse array of compounds will generate PAP, a potent inhibitor of the sulphotransferases itself. Therefore, we propose that PAP phosphatases are necessary in plants for the function of sulphotransferases, and not for reductive sulphate assimilation as in yeast.

4. An animal lithium-sensitive PAP phosphatase

The sulphation reactions in mammals affect many different physiological processes, including deactivation and bioactivation of xenobiotics; inactivation of hormones and catecholamines; structure and function of macromolecules, and elimination of end products of catabolism (Klaassen and Boles, 1997). Sulphation involves the transfer of a sulphate group from PAPS to an acceptor molecule in a reaction that is catalysed by a family of sulphotransferase enzymes (Weinshilboum *et al.*, 1997). The importance of sulphation in animals is illustrated by the phenotype of brachymorphic mice, which lack sulphurylase kinase activity, the bifunctional enzyme that synthesizes PAPS.

Brachymorphic mice have abnormal hepatic detoxification, bleeding times and post-natal growth.

As PAP is an end product in any sulphation reaction and a multitude of sulpho-conjugates exist in animals, it is reasonable to predict the existence of animal PAP phosphatases. We have confirmed this prediction by cloning the first mammalian PAP phosphatase, RnPIP, through functional complementation of the yeast *hal2* mutant with a cDNA library from *Rattus norvegicus* (Lopez-Coronado *et al.*, 1999). Biochemical characterization of the recombinant protein confirms that RnPIP converts PAP to AMP in a magnesium-dependent reaction which is very sensitive to inhibition by lithium (*Table 1*). However, RnPIP is not sodium-sensitive since animal enzymes are adapted to the high intracellular sodium levels (20–50 mM) present in animal cells. Database searches confirm the presence of ESTs (expressed sequence tags) that correspond to the human homologue of RnPIP. We predict that the human homologue of RnPIP will also be lithium-sensitive. Therefore, this enzyme is the third animal enzyme to be inhibited by Li⁺, the other two being inositol monophosphatase and inositol polyphosphate 1-phosphatase. The mechanism of lithium therapy for treatment of bipolar disorder has been assumed to consist of inhibition of inositol monophosphatases, resulting in a run-down of the inositol cycle of calcium signalling in the brain (Berridge and Irvine, 1989). The recent identification of a new lithium-sensitive enzyme provides another possible target in the mechanism of action of lithium therapy.

5. Concluding remarks

The study of the *HAL2* gene in the model organism *S. cerevisiae* paved the way for the identification of *HAL2*-like genes in higher organisms. Plant and animal *HAL2*-like genes have been cloned mainly through functional complementation of the auxotrophy for methionine of *hal2* cells. As a result, a sodium-sensitive PAP phosphatase has been identified in *Arabidopsis*, which could represent a potential target of salt toxicity in plants. The mammalian PAP phosphatase is a lithium-sensitive enzyme and it constitutes a new target in lithium treatment, which might contribute to the therapeutic or toxic effects observed in patients treated with lithium. The elucidation of the three-dimensional structure of this family of enzymes will improve our understanding of their inhibition by lithium and sodium, and might facilitate the engineering of new versions of the enzymes with a modified sensitivity to these cations.

References

Berridge, M.J. and Irvine, R.F. (1989) Inositol phosphates and cell signalling. *Nature* **341**: 197–205.
Dichtl, B., Stevens, A. and Tollervey, D. (1997) Lithium toxicity in yeast is due to the inhibition of RNA processing enzymes. *EMBO J.* **16**: 7184–7195.
Faulkner, I.J. and Rubery, P.H. (1992) Flavonoids and flavonoid sulphates as probes of auxin-transport regulation in *Cucurbita pepo* hypocotyl segments and vesicles. *Planta* **186**: 618–625.
Gil-Mascarell, R., Lopez-Coronado, J.M., Belles, J.M., Serrano, R. and Rodriguez, P.L. (1999) The *Arabidopsis HAL2*-like gene family includes a novel sodium-sensitive phosphatase. *Plant J.* **17**: 373–383.

Glaser, H.-U., Thomas, D., Gaxiola, R., Montrichard, F., Surdin-Kerjan, Y. and Serrano, R. (1993) Salt tolerance and methionine biosynthesis in *Saccharomyces cerevisiae* involve a putative phosphatase gene. *EMBO J.* **12:** 3105–3110.

Glendening, T.M. and Poulton, J.E. (1990) Partial purification and characterization of a 3′-phosphoadenosine 3′-phosphosulfate: desulfoglucosinolate sulfotransferase from cress (*Lepidium satirum*). *Plant Physiol.* **94:** 811–818.

Gutierrez-Marcos, J.F., Roberts, M.A., Campbell, E.I. and Wray, J.L. (1996) Three members of a novel small gene-family from *Arabidopsis thaliana* able to complement functionally an *Escherichia coli* mutant defective in PAPS reductase activity encode proteins with a thioredoxin-like domain and 'APS reductase' activity. *Proc. Natl Acad. Sci. USA* **93:** 13377–13382.

Hanson, A.D., Rathinasabapathi, B., Rivoal, J., Burnet, M., Dillon, M.O. and Gage, D.A. (1994) Osmoprotective compounds in the *Plumbaginaceae*: a natural experiment in metabolic engineering of stress tolerance. *Proc. Natl Acad. Sci. USA* **91:** 306–310.

Jain, J.C., GrootWassink, J.W.D., Kolenovsky, A.D. and Underhill, E.W. (1990) Purification and properties of a 3′-phosphoadenosine 3′-phosphosulfate: desulfoglucosinolate sulfotransferase from *Brassica juncea* cell cultures. *Phytochem.* **29:** 1425–1428.

Jones, E.W. and Fink, G.R. (1982) Amino acid and nucleotide biosynthesis. In: *The Molecular Biology of the Yeast Saccharomyces. Metabolism and Gene Expression* (eds J.N. Strathern, E.W. Jones and J.R. Broach). Cold Spring Harbor Laboratory, Cold Spring Harbor, NY, pp. 221–230.

Klaasen, C.D. and Boles, J.W. (1997) The importance of 3′-phosphoadenosine 5′-phosphosulfate (PAPS) in the regulation of sulfation. *FASEB J.* **11:** 404–418.

Lopez-Coronado, J.M., Belles, J.M., Lesage, F., Serrano, R. and Rodriguez, P.L. (1999) A novel mammalian lithium-sensitive enzyme with a dual enzymatic activity: 3′-phosphoadenosine 5′-phosphate (PAP) phosphatase and inositol polyphosphate 1-phosphatase. *J. Biol. Chem.* (in press).

Mithen, R.F. (1992) Leaf glucosinolate profiles and their relationship to pest and disease resistance in oilseed rape. *Euphytica* **63,** 71–83.

Murguia, J.R., Belles, J.M. and Serrano, R. (1995) A salt-sensitive 3′(2′),5′-bisphosphate nucleotidase involved in sulfate activation. *Science* **267:** 232–234.

Murguia, J.R., Belles, J.M. and Serrano, R. (1996) The yeast HAL2 nucleotidase is an *in vivo* target of salt toxicity. *J. Biol. Chem.* **271:** 29029–29033.

Quintero, F.J., Garciadeblas, B. and Rodriguez-Navarro, A. (1996) The *SAL1* gene of *Arabidopsis*, encoding an enzyme with 3′(2′),5′-bisphosphate nucleotidase and inositol polyphosphate 1-phosphatase activities, increases salt tolerance in yeast. *Plant Cell* **8:** 529–537.

Roth, J.A., Rivett, A.J. and Renskers, K.J. (1982) Properties of human brain phenolsulfotransferase and its role in the inactivation of catecholamine neurotransmitters. In: *Sulfate Metabolism and Sulfate Conjugation* (eds G.J. Mulder, J. Caldwell, G.M.J. Van Kemper and R.J. Vonk). Taylor and Francis, London, pp. 107–114.

Schwenn, J.D. and Schriek, U. (1987) PAPS-reductase from *Escherichia coli*: characterization of the enzyme as probe for thioredoxins. *Z. Naturforschung* **42:** 93–102.

Schwenn, J.D., Krone, F.A. and Husmann, K. (1988) Yeast PAPS reductase: properties and requirements of the purified enzyme. *Arch. Microbiol.* **150:** 313–319.

Setya, A., Murillo, M. and Leustek, T. (1996) Sulfate reduction in higher plants: molecular evidence for a novel 5′-adenylylsulfate reductase. *Proc. Natl Acad. Sci. USA* **93:** 13383–13388.

Thomas, D., Barbey, R., Henry, D. and Surdin-Kerjan, Y. (1992) Physiological analysis of mutants of *Saccharomyces cerevisiae* impaired in sulphate assimilation. *J. Gen. Microbiol.* **138:** 2021–2028.

Varin, L. and Ibrahim, R.K. (1992) Novel flavonol 3-sulfotransferase. *J. Biol. Chem.* **267,** 1858–1863.

Varin, L., Marsolais, F., Richard, M. and Rouleau, M. (1997a) Biochemistry and molecular biology of plant sulfotransferases. *FASEB J.* **11:** 517–525.

Varin, L., Chamberland, H., Lafontaine, J.G. and Richard, M. (1997b) The enzyme involved in sulfation of the turgorin, gallic acid 4-*O*-(β-D-glucopyranosyl-6′-sulfate) is pulvini-localized in *Mimosa pudica. Plant J.* **12:** 831–837.

Weinshilboum, R.M., Otterness, D.M., Aksoy, I.A., Wood, T.C., Her, C. and Raftogianis, R.B. (1997) Sulfotransferase molecular biology: cDNAs and genes. *FASEB J.* **11:** 3–14.

Chapter 23

Studies of salt stress- and water stress-regulated genes: the stress-regulated *Asr1* gene encodes a DNA-binding protein

Dudy Bar-Zvi and Ayelet Gilad

1. Background

Water is the most abundant compound in living organisms and makes up 70–90% of the weight of most forms of life. Water also represents the continuous phase of living organisms. It is thus clear why water availability is a major factor in the distribution of species, and in maintaining life. The response of plants to water deficit involves a number of metabolic and physiological changes, many of which have not been fully characterized. These responses to water stress include physiological changes, such as stomatal closure, decreased rates of photosynthesis and decreased photorespiration, accumulation of small organic molecules, alterations in plant hormone levels, and changes in gene expression (for recent reviews see Bartels and Nelson, 1994; Bonhert *et al.*, 1995; Bray, 1997; Shinozaki and Yamaguchi-Shinozaki, 1996; Yamaguchi-Shinozaki and Shinozaki, 1997). Abscisic acid (ABA) is recognized as the major plant hormone involved in the response to salt and osmotic stresses: during stress, ABA concentration increases dramatically in all plant organs (see Giraudat *et al.*, 1994; Zeevaart, 1988 for reviews). Furthermore, ABA application to plants elicits similar responses to those resulting from water stress.

Plant genes can be classified into three groups according to their response to stress, depending on whether their expression increases, decreases or whether it is not affected by the stress. Most genes that have been studied belong to the group whose expression is increased by stress. In kinetic terms, these genes can be further divided into two groups: one in which the mRNA level increases and stays high until the stress is relieved and the other in which the increase in mRNA level is transient.

Stress-regulated genes vary in their tissue specificity. Some are root or leaf specific while others are expressed in all plant organs. Some genes are characterized by low

Plant Responses to Environmental Stress, edited by M.F. Smallwood, C.M. Calvert and D.J. Bowles.
© 1999 BIOS Scientific Publishers, Oxford.

levels of expression in all the organs of unstressed plants, which becomes elevated following stress, whereas in other genes, mRNA is detected only following stress.

Genes induced by application of ABA, by drought or salt stress are usually believed to be regulated at the transcriptional level. A few promoters of these genes have been isolated and studied, and *cis*-acting elements have been functionally identified and characterized (for recent reviews see Bartels and Nelson, 1994; Bonhert *et al.*, 1995; Bray, 1997; Shinozaki and Yamaguchi-Shinozaki, 1996; Yamaguchi-Shinozaki and Shinozaki, 1997). Moreover, good correlation between mRNA and protein levels has been shown for some genes (e.g. dsp22 from *Craterostigma plantagineum*, Bartels *et al.*, 1992; *HVA1* from barley, Hong *et al.*, 1992). However, in other cases there is a marked delay in protein induction compared with mRNA accumulation (e.g. rice *SalT*, Claes *et al.*, 1990). On the other hand, changes in mRNA levels are not always followed by similar changes in the corresponding protein (e.g. tobacco osmotin, LaRosa *et al.*, 1992). Moreover, the *Brassica* BnD22 protein was translated during water stress, although its mRNA was present in both water-stressed and non-stressed plants (Reviron *et al.*, 1992). The latter cases clearly implicate additional regulation at the translation level.

2. Biological function of proteins encoded by water stress-/abscisic acid-regulated genes

The plant response to salt and water abiotic stress is complex and involves a large number of events. Proteins involved in the plant response to stress can be classified into two groups, those that take part in the signal-transduction pathways, and those that are end products of these molecular events. Proteins of the first group described above include DNA- and RNA-binding proteins, as well as protein kinases. The second group contains proteins that are involved in stress response and includes enzymes responsible for the synthesis of compatible solutes and scavenging of reactive oxygen species, water channels, ion transporters and protective proteins.

Transgenic plants over-expressing stress-regulated genes have been produced (for review see Holmberg and Bulow, 1998). The genes introduced encode some of the enzymes mentioned above, or proteins that are accumulated in response to stress whose role is not fully understood, for example late embryogenesis abundant (LEA) proteins. These transgenic plants turned out to be only slightly more tolerant than wild-type plants. Therefore, it will be necessary to produce transgenic plants expressing simultaneously several genes in order to achieve a better tolerance. This goal could be obtained by introducing multiple genes into a single plant, or by manipulation of a master gene that functions in the signal-transduction pathway and regulates a battery of genes.

3. Tomato *Asr1* gene and protein

Asr1 was cloned by differential screening of a tomato fruit cDNA library with cDNA derived from irrigated versus water-stressed tomato leaves (Iusem *et al.*, 1993). *Asr1* encodes a hydrophilic 13 kDa polypeptide, which has an unusual amino acid composition

(a)

```
  1 MEEEKHHHHHLFHHKDKAEE
 21 GPVDYEKEIKHHKHLEQIGK
 41 LGTVAAGAYALHEKHEAKKD
 61 PEHAHKHKIEEEIAAAAAVG
 81 AGGFAFHEHHEKKDAKKEEK
101 KKLRGDTTISSKLLF
```

(b)

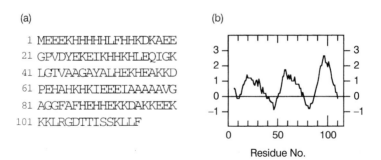

Residue No.

Figure 1. (a) Amino acid sequence and (b) Kyte and Doolittle hydropathy plot of ASR1 protein.

(*Figure 1*). Four amino acid residues (lysine, glutamic acid, histidine and alanine) contribute 62% of the total residues of the polypeptide. The protein is highly charged, it contains 23 residues of acidic amino acids, 21 of basic amino acids and 18 histidine residues. Interestingly, there is a strong preference for acidic or basic amino acids; glutamic acid and lysine are 18 and 20 residues, respectively, whereas there are only five and one residues of aspartic acid and arginine, respectively. The ASR1 protein lacks cysteine, asparagine and tryptophan.

In tomato, *Asr1* is a member of a small gene family. Four genes were cloned (Amitai-Zeigerson *et al.*, 1994; Gilad *et al.*, 1997; Rossi and Iusem, 1994). In tomato, the *Asr1*, *Asr2* and *Asr3* genes were mapped to be on closely linked loci on chromosome 4 (Rossi *et al.*, 1996). ASR1 protein is not specific to tomato or plants from the solanaceae family; anti-ASR1 antiserum reacted with proteins in plants from dicots, monocots and gymnosperms (D. Bar-Zvi, unpublished data). Accordingly, *Asr1* homologues were cloned from several plant species (*Table 1*).

The effects of water stress, salt stress and ABA on the steady-state levels of tomato *Asr1* mRNA and protein were studied (Amitai-Zeigerson *et al.*, 1995). Low levels of mRNA and protein were detected in the roots and shoots of hydroponically grown tomato plants. Application of polyethylene glycol (PEG), NaCl or ABA to the growth medium resulted in an elevation of the steady-state levels of both *Asr1* mRNA and protein (Amitai-Zeigerson *et al.*, 1995). This increase was transient, reaching a maximum approximately 24 h after the application of the stress. The extent of the increase correlated with the severity of the stress (Amitai-Zeigerson *et al.*, 1995). The similarity between the response of mRNA and protein to the stress suggests that the *Asr1* gene is regulated mainly by RNA transcription or RNA stability.

Anti-ASR1 antiserum was used to screen the cellular location of the ASR1 protein (Gilad *et al.*, 1997). ASR1 protein was detected in the nuclear fraction. Treatment of the nuclei with triton X-100 did not solubilize the protein, suggesting that the protein is bound to chromatin (Gilad *et al.*, 1997).

Asr1 was cloned into pT7-2 expression vector and was expressed in *Escherichia coli* Bl-21 cells (Gilad *et al.*, 1997). The expressed ASR1 protein was soluble and was purified to homogeneity by chromatography on heparin-agarose followed by chromatography on nickel-agarose (Gilad *et al.*, 1997). Nickel-agarose is widely used for the

Table 1. Cloned plant genes homologous to tomato *Asr1*

Plant species	Accession number
Lycopersicon esculentum	L08255, U86130, X72733 (*Asr1*) X74907, L20756 (*Asr2*) X74908 (*Asr3*)
Citrus maxima	U18972
Mesembryanthemum crysatallinum	AF054443, AF053566
Oryza sativa	AF039573, AU029694, AU029904
Pinus taeda	U59424, U59451, U67135, U52865
Prunus armenica	U93164
Solanum chaoense	U12439
Zea mays	U09276

Asr1 homologues were searched in Genebank and EMBL databases using FASTA and BLAST programs.

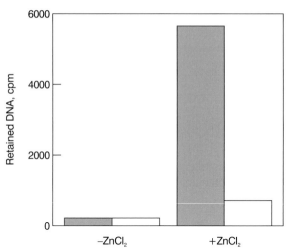

Figure 2. ASR1 has a zinc-dependent DNA-binding activity. ASR1 protein was expressed in *Escherichia coli* and purified to homogeneity. Radiolabelled DNA incubated with ASR1 (shaded) or without added proteins (open) in a reaction mix containing 0 or 0.8 mM $ZnCl_2$. Mixes were filtered through nitrocellulose using dot-blot apparatus. Wells were washed with the binding buffer. Radioactivity retained on the filter was determined by a scintillation counter.

purification of polyhistidine-tagged proteins. The native ASR1 protein has a penta histidine sequence at residues 6–10 (*Figure 1a*), which enabled us to use this approach to purify native ASR1. The purified ASR1 protein was used for DNA-binding assays. DNA binding by ASR1 was dependent on Zn^{2+} (*Figure 2*). Many DNA-binding proteins were shown to contain bound zinc. The Zn^{2+} ion may be co-ordinated to residues of histidine, glutamic acid, aspartic acid or cysteine (see Vallee and Auld, 1992, for review). ASR1 contains 18, 18, 5 and 0 residues of these amino acids, respectively. In contrast to

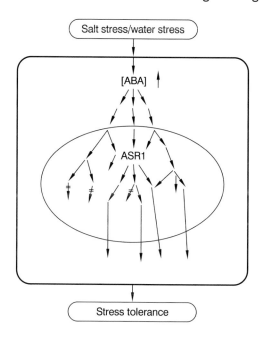

Figure 3. Proposed model of the role of ASR1 in stress response. ASR1 is a DNA-binding protein; ABA is abscisic acid. The outer and the inner lines represent the cell and nucleus, respectively.

zinc-enzymes, virtually all DNA-binding proteins contain multiple zinc atoms (Vallee and Auld, 1992). Three types of zinc-binding sites in DNA-binding proteins were proposed (Vallee and Auld, 1992). These motifs are termed zinc-finger, zinc-twist and zinc-cluster. Amino acid sequence analysis revealed that none of these motifs are present in ASR1. Furthermore, the Zn^{2+} ions in DNA-binding proteins are co-ordinated primarily to cysteine residues (and to some extent to histidine residues), that are missing in ASR1. Thus, ASR1 might belong to a new class of zinc-binding DNA-binding proteins.

Our hypothesis for the role of ASR1 protein in plant stress response is presented in *Figure 3*. The regulation of gene expression by water-stress is complex and involves several routes, some are ABA-dependent whereas others are ABA-independent. ASR1 might function in regulation of genes in ABA-dependent pathway(s). ASR1 activity might result in the activation or repression of gene expression. The interaction of ASR1 with DNA, zinc and its effect on the expression of other genes is being investigated.

Acknowledgements

This work was supported by a grant from the Basic Research Foundation of The Israel Academy of Sciences and Humanities to D.B.Z.

References

Amitai-Zeigerson, H., Scolnik, P.A. and Bar-Zvi, D. (1994) Genomic nucleotide sequence of tomato *Asr2,* a second member of the stress/ripening-induced *Asr1* gene family. *Plant Physiol.* **106:** 1699–1700.

Amitai-Zeigerson, H., Scolnik, P.A. and Bar-Zvi, D. (1995) Tomato *Asr1* mRNA and protein are transiently expressed following salt stress, osmotic stress and treatment with abscisic acid. *Plant Sci.* **110:** 205–213.

Bartels, D. and Nelson, D. (1994) Approaches to improve stress tolerance using molecular studies. *Plant Cell Environ.* **17:** 659–667.

Bartels, D., Hanke, C., Schneider, K., Michel, D. and Salamini, F. (1992) A desiccation-related Elip-like gene from the resurrection plant *Craterostigma plantagineum* is regulated by light and ABA. *EMBO J.* **11:** 2771–2778.

Bonhert, H.J., Nelson, D.E. and Jensen, R.G. (1995) Adaptation to environmental stresses. *Plant Cell* **7:** 1099–1111.

Bray, A.E. (1997) Plant responses to water deficit. *Trends Plant Sci.* **2:** 48–54.

Claes, B., Dekeyser, R., Villarroel, R., Van den Bulcke, M., Bauw, G., Van Montagu, M. and Caplan, A. (1990) Characterization of a rice gene showing organ-specific expression in response to salt stress and drought. *Plant Cell* **2:** 19–27.

Gilad, A., Amitai-Zeigerson, H., Scolnik, P.A. and Bar-Zvi, D. (1997) *Asr1*, a tomato water-stress regulated gene: genomic organization, developmental regulation and DNA-binding activity. *Acta Hort.* **447:** 447–453.

Giraudat, J., Parcy, F., Bertauche. N., Gosti, F., Leung, J., Morris, P.C., Bouvier-Durand, M. and Vartanian, N. (1994) Current advances in abscisic acid action and signalling. *Plant Mol. Biol.* **26:** 1557–1577.

Holmberg, N. and Bulow, L. (1998) Improving stress tolerance in plants by gene transfer. *Trends Plant Sci.* **3:** 61–66.

Hong, B., Barg, R. and Ho, T.D. (1992) Developmental and organ-specific expression of an ABA- and stress-induced protein in barley. *Plant Mol. Biol.* **18:** 663–674.

Iusem, N.D., Bartholomew, D.M., Hitz, W.D. and Scolnik, P.A. (1993) Tomato (*Lycopersicon esculentum*) transcript induced by water deficit and ripening. *Plant Physiol.* **102:** 1353–1354.

LaRosa, P.C., Chen, Z., Nelson, D.E., Singh, N.K., Hasegawa, P.M. and Bressan, R.A. (1992) Osmotin gene expression is posttranscriptionally regulated. *Plant Physiol.* **100:** 409–415.

Reviron, M.P., Vartanian, N., Sallantin, M., Huet, J.C., Pernollet, J.C. and de Vienne, D. (1992) Characterization of a novel protein induced by progressive or rapid drought and salinity in *Brassica napus* leaves. *Plant Physiol.* **100:** 1486–1493.

Rossi, M. and Iusem, N.D. (1994) Tomato (*Lycopersicon esculentum*) genomic clone homologous to a gene encoding an abscisic acid-induced protein. *Plant Physiol.* **104:** 1073–1074.

Rossi, M., Lijavetzky, D., Bernacchi, D., Hopp, H.E. and Iusem, N. (1996) *Asr* genes belong to a gene family comprising at least three closely linked loci on chromosome 4 in tomato. *Mol. Gen. Genet.* **252:** 489–492.

Shinozaki, K. and Yamaguchi-Shinozaki, K. (1996) Molecular responses to drought and cold stress. *Curr. Opinion Biotechnol.* **7:** 161–167.

Vallee, B.L. and Auld, D.S. (1992) Functional zinc-binding motifs in enzymes and DNA-binding proteins. *Faraday Discuss.* **93:** 47–65.

Yamaguchi-Shinozaki, K. and Shinozaki, K. (1997) Gene expression and signal transduction in water-stress response. *Plant Physiol.* **115:** 327–334.

Zeevaart, J.A.D. (1988) Metabolism and physiology of abscisic acid. *Annu. Rev. Plant Physiol. Plant Mol. Biol.* **39:** 439–473.

Chapter 24

Salt-induced proteins related to oxidative stress: PHGPX and other proteins of the Halliwell–Asada cycle

Gozal Ben-Hayyim, Doron Holland and Yuval Eshdat

1. Background

We have previously isolated a salt stress-associated protein from citrus, as well as its encoding gene (Ben-Hayyim *et al.*; 1993; Holland *et al.*, 1993). This protein was demonstrated to be a phospholipid hydroperoxide glutathione peroxidase (Cit-PHGPX), which had not been identified before in plants (Holland *et al.*, 1994; Beeor-Tzahar *et al.*, 1995). The plant enzyme differs from its animal analogues by carrying a cysteine amino acid side chain as its catalytic residue, instead of a selenocysteine (Faltin *et al.*, 1998), which probably explains its lower catalytic activity towards lipid hydroperoxides. Following this finding, that salt stress in citrus induces the expression of a plant peroxidase, we have further studied the relationship between salt and oxidative stress in plants, in an attempt to learn about the biological role of Cit-PHGPX and whether it integrates in the present model of how plants confront oxidative and salt stress. In our current study, we have focused on several key questions, the answers to which may shed light on the molecular mechanism that crosslinks these two effects:

(i) Does high salt concentration directly induce *csa*, the gene encoding the Cit-PHGPX, or does salt stress lead to formation of other inducers of this gene?
(ii) What regulates the gene expression at the levels of both mRNA transcript and protein production?
(iii) What is the biological function of PHGPX in plants and does it confer increase in tolerance, either directly or indirectly, towards abiotic stress?

We are now at an advanced stage of isolation of the promoter of *csa*. We are also searching for a potential inducer for this gene. Kinetic studies of the gene expression products under various conditions of salt-sensitive and salt-tolerant citrus cell lines, exposed to various external factors such as heat, cold, NaCl, abscisic acid (ABA) and

---

Plant Responses to Environmental Stress, edited by M.F. Smallwood, C.M. Calvert and D.J. Bowles.

alkyl hydroperoxides, have been carried out. Such experiments will help us to understand better the relationship between the salt stress and the expression of the antioxidative enzyme, PHGPX. In fact, the results we have obtained so far already indicate that
salt stress in plants probably leads to the formation of specific oxyradicals which mediate the induction of *csa*, and to the expression of the plant PHGPX. The expression of
this enzyme under salt stress raises questions as to the role of the plant antioxidant
defence mechanism in response to the stress, and the specific role played by PHGPX in
this system.

2.   Proteins of the Halliwell–Asada cycle

Oxygen radicals are generated during plant metabolism and they should be scavenged
for maintenance of normal growth. In the ascorbate–glutathione, also named the
Halliwell–Asada cycle (Asada, 1994), two key enzymes are assumed to deal with the
detoxification of reactive oxygen species (ROS) in plants. The primary scavenger is the
enzyme superoxide dismutase (SOD) which converts superoxide to hydrogen peroxide.
This toxic product of SOD is eliminated by ascorbate peroxidase (APX) at the expense
of oxidizing ascorbate to monodehydroascorbate, which in turn is recycled to ascorbate
by glutathione reductase. These enzymes exist as several isozymes and are active in the
chloroplasts and the cytosol (Asada, 1994).

A large body of evidence has accumulated from various plant systems showing that
environmental stresses, especially drought and salt stress, alter the amounts and the
activities of enzymes involved in scavenging oxygen radicals and their corresponding
steady-state level of mRNA. Activities of cytosolic and chloroplastic Cu/Zn-SOD
isozymes and cytosolic APX, as well as their corresponding mRNA transcripts, were
increased by drought treatment of pea plants. Similarly, osmotic stress induced the
increase of Mn-SOD transcript abundance in maize. The mRNA transcript corresponding to this isozyme, as well as the one for Cu/Zn-SOD, was also induced by chilling of
tobacco plants. In addition, increased activities of the respective mitochondrial and
chloroplastic SOD and APX isozymes were observed in pea plants under salt stress (for
related references see Eshdat *et al.*, 1997).

Contradictory results were obtained in relation to the effect of salt and drought on the
activity and protein levels of the various isozymes of SOD and APX (see discussion in
Gueta-Dahan *et al.*, 1997). Salt was shown to induce the expression of a protein having
a strong homology to APX in radish, while in leaves of *Vigna unguiculata*, cytosolic
APX was slightly reduced and chloroplastic APX unchanged. It has been reported that
in NaCl-tolerant pea cultivars, leaf mitochondrial Mn-SOD and chloroplast Cu/Zn-SOD
activities increased under salt stress, while the cytosolic Cu/Zn-SOD activity remained
unchanged (Hernandez *et al.*, 1993b, 1995). In the salt-sensitive cultivar neither APX
nor chloroplastic SOD was increased by salt, while the cytosolic and mitochondrial
SOD even decreased. In salt-sensitive and salt-tolerant cultured citrus cells and leaf tissues it was shown that only cytosolic Cu/Zn-SOD activity was increased by salt
whereas the activity of other isoforms has not been changed (Gueta-Dahan *et al.*, 1997).
It was also demonstrated that APX activity was reduced by salt, and that its constitutive

activity was much higher in salt-tolerant cells. It is suggested that under salt stress excess hydrogen peroxide is formed leading to a series of reactions producing lipid hydroperoxides, which in turn induce the expression of PHGPX (see above).

It is questionable whether one can integrate these contradictory data into a model and draw general conclusions. The experiments exploit different plant systems, with cultivars varying with respect to salt tolerance and the effects were measured in different tissues and at various ages. Data concerning the various SOD and APX isozymes show that a decrease of one isozyme is either accompanied by a decrease of the others, or by no significant change (Hernandez et al., 1993b, 1995). No data could be found showing a salt-induced increase in one isozyme and a decrease in another. One can assume that under conditions where inhibition of activity is found, the tissue is under severe stress where total metabolism is suppressed. On the other hand, a stress-induced enzyme activity may indicate a specific role in coping with the stress. Thus, it is not surprising that an increase, or a slower loss, of activities of SOD and APX was found in salt- and drought-tolerant cultivars of pea, sorghum, cotton, soybean and maize (Chen et al., 1997; Del Longo et al., 1993; Gossett et al., 1994; Hernandez et al., 1993b, 1995; Jagtap and Bhargava, 1995). It should be noted that salt induces an increase in the cytosolic Cu/Zn-SOD both in Lupinus and citrus (Gueta-Dahan et al., 1997; Przymusinski et al., 1995), and acclimation to water stress resulted in concentration of antioxidant enzymes in the cytosol (Huang and Chen, 1995).

In recent years, a reasonable amount of data indicates that transformed plants over-expressing various isoforms of SOD have improved tolerance to oxidative stress imposed by light and methyl viologen. The degree of enhanced tolerance varies with the external conditions and it seems that it does not require the expression of the specific isoform at its respective organelle (see discussion of Slooten et al., 1995). Only one report demonstrates that transgenic alfalfa plants over-expressing Mn-SOD were more tolerant to water deficit conditions (McKersie et al., 1996). Growth under saline conditions was not determined. However, tobacco plants over-expressing Fe-SOD did not show improved growth under salt stress, or under a variety of other stresses, for example oxidative, cold or high light (Van Camp et al., 1996). It is suggested that unlike the wild-type tobacco, the cytosolic Cu/Zn-SOD is not induced in the transgenic plants, and its contribution to salt tolerance might be crucial.

3. Other proteins

It is noteworthy that enzymes producing H_2O_2 are induced by salt. In addition to SOD two other proteins, namely oxalate oxidase and cytochrome c oxidase were shown to be induced by salt treatment. The first protein was identified from a salt-induced clone in barley roots, and was found to be homologous to germin, a protein independently identified as oxalate oxidase (Hurkman and Tanaka, 1996). The activity of the second enzyme was increased in salt-treated Vigna plants (Hernandez et al., 1993a). It is possible to speculate that H_2O_2 has a role in signal transduction of the stress. On the other hand, the level of several proteins possessing activity in reducing activity were shown to be increased by salt. In addition to APX and PHGPX discussed above, salt increased the activity and gene expression of peroxidases, glutathione-S-transferase (GST), aldose reductase and others, and tobacco plants transformed with active GST/PHGPX showed increased tolerance to salinity, whereas catalase-deficient transgenic tobacco plants showed higher sensitivity to the stress (Willekens *et al.*, 1997).

4. Concluding remarks

There are indications that maintenance of a more reduced state sustains salt tolerance. This is in agreement with results showing correlation between salt tolerance and higher constitutive activities of catalase in cotton (Banks *et al.*, 1997), and APX in citrus (Gueta-Dahan *et al.*, 1997), and from results obtained with transgenic plants. Our recent results indicate that salt induction of PHGPX is retarded in salt-tolerant cells vs. salt-sensitive ones, and is similarly and directly induced by hydroperoxides in both types of cell (O. Avsian-Kretchmer, Y. Eshdat, Y. Gueta-Dahan and G. Ben-Hayyim, unpublished data). In addition, overloading cells with antioxidants abolished the salt-induced increase of *csa*, the gene encoding PHGPX in citrus. These results, and others, led us to suggest that ROS are produced by salt as an early event in the regulation of *csa* expression (*Figure 1*). Hydroperoxides, probably of a lipid type, are formed when the scavenging capacity of ROS is poor, leading to PHGPX induction (Gueta-Dahan *et al.*, 1997).

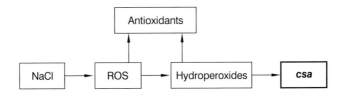

Figure 1. A suggested scheme for the regulation of *csa* expression. ROS, reactive oxygen species.

References

Asada, K. (1994) Production and action of active oxygen in photosynthetic tissues. In: *Causes of Photooxidative* (eds C.H. Foyer and P.M. Mullineaux). CRC Press, Boca Raton, FL, pp. 77–104.

Banks, S.W., Rajguru, S.N., Gossett, D.R. and Millhollon, E.P. (1997) Antioxidant response to salt stress during fiber development. In: *Proceedings of the Beltwide Cotton Conferences*, vol. 2. National Cotton Council of America, Memphis, TN, pp. 1422–1426.

Beeor-Tzahar, T., Ben-Hayyim, G., Holland, D., Faltin, Z. and Eshdat, Y. (1995) A stress-associated citrus protein is a distinct plant phospholipid hydroperoxide glutathione peroxidase. *FEBS Lett.* **366**: 151–155.

Ben-Hayyim, G., Faltin, Z., Gepstein, S., Camoin, L., Strosberg, A.D. and Eshdat, Y. (1993) Isolation and characterization of salt-associated protein in citrus. *Plant Sci.* **88**: 129–140.

Chen, Y.W., Shao, G.H. and Chang, R.Z. (1997) The effect of salt stress on superoxide dismutase in various organelles of cotyledons of soybean seedlings. *Acta Agron. Sinica* **23**: 214–219.

Del Longo, O.T., Gonzalez, C.A., Pastori, G.M. and Trippi, V.S. (1993) Antioxidant defences under hyperoxygenic hyperosmotic conditions in leaves of two lines of maize with differential sensitivity to drought. *Plant Cell Physiol.* **34**: 1023–1028.

Eshdat, Y., Holland, D., Faltin, Z. and Ben-Hayyim, G. (1997) Plant glutathione peroxidases. *Physiol. Plant.* **100**: 234–240

Faltin, Z., Ben-Hayyim, G., Camoin, L., Perl, A., Beeor-Tzahar, T., Strosberg, A.D., Holland, D. and Eshdat, Y. (1998) Non-selenium plant phospholipid hydroperoxide glutathione peroxidase is over-expressed under salt and oxidative stress. *Physiol. Plant* **104**: 741–746.

Gossett, D.R., Millhollon, E.P., Lucas, M.C., Banks, S.W. and Marney, M.M. (1994) The effects of NaCl on antioxidant enzyme activities in callus tissue of salt-tolerant and salt-sensitive cotton cultivars (*Gossypium hirsutum* L.). *Plant Cell Rep.* **13**: 498–503.

Gueta-Dahan, Y., Yaniv, Z., Zilinskas, B.A. and Ben-Hayyim, G. (1997) Salt and oxidative stress: similar and specific responses and their relation to salt tolerance in citrus. *Planta* **203**: 460–469.

Hernandez, J.A., Almansa, A.S., del Rio, A. and Sevilla, F. (1993a) The effects of salinity on metalloenzymes of oxygen metabolism in two leguminous plants. *J. Plant Nutr.* **16**: 2539–2554.

Hernandez, J.A., Corpas, F.J., Gomez, M., del Rio, L.A. and Sevilla, F. (1993b) Salt-induced oxidative stress mediated by activated oxygen species in pea leaf mitochondria. *Physiol. Plant.* **89**: 103–110.

Hernandez, J.A., Olmos, E., Corpas, F.J., Sevilla, F. and del Rio, L.A. (1995) Salt-induced oxidative stress in chloroplasts of pea plants. *Plant Sci.* **105**: 151–167

Holland, D., Ben-Hayyim, G., Faltin, Z., Camoin, L., Strosberg, A.D. and Eshdat, Y. (1993) Molecular characterization of salt-stress associated protein in citrus: protein and cDNA sequence homology to mammalian glutathione peroxidases. *Plant Mol. Biol.* **21**: 923–927.

Holland, D., Faltin, Z., Perl, A., Ben-Hayyim, G. and Eshdat, Y. (1994) A novel plant glutathione peroxidase-like protein provides tolerance to oxygen radicals generated by paraquat in *E. coli.* *FEBS Lett.* **337**: 52–55.

Huang, C.Y. and Chen, Y.M. (1995) Role of glutathione reductase and related enzymes in salt-tolerant mechanism of soybean plants grown under salt-stress conditions. *Taiwania* **40**: 21–34.

Hurkman, W.J. and Tanaka, C.K. (1996) Effect of salt stress on germin gene expression in barley roots. *Plant Physiol.* **110**: 971–977.

Jagtap, V. and Bhargava, S. (1995) Variation in the antioxidant metabolism of drought tolerant and drought susceptible varieties of *Sorghum bicolor* (L.) Moench. exposed to high light, low water and high temperature stress. *J. Plant Physiol.* **145**: 195–197.

McKersie, B.D., Bowley, S.R., Harjanto, E. and Leprince, O. (1996) Water-deficit tolerance and field performance of transgenic alfalfa overexpressing superoxide dismutase. *Plant Physiol.* **111**: 1177–1181.

Przymusinski, R., Rucinska, R. and Gwozdz, E.A. (1995) The stress-stimulated 16 kDa polypeptide from lupin roots has properties of cytosolic Cu:Zn-superoxide dismutase. *Environ. Exp. Bot.* **35**: 485–495.

Slooten, L., Capiau, K., Van Camp, W., Van Montagu, M., Sybesma, C. and Inze, D. (1995) Factors affecting the enhancement of oxidative stress tolerance in transgenic tobacco overexpressing manganese superoxide dismutase in the chloroplasts. *Plant Physiol.* **107**: 737–750.

Van Camp, W., Capiau, K., van Montagu, M., Inze, D. and Slooten, L. (1996). Enhancement of oxidative stress tolerance in transgenic tobacco plants overproducing Fe-superoxide dismutase in chloroplasts. *Plant Physiol.* **112**: 1703–1714.

Willkens, H., Chamnogpol, S., Schraudner, M., Langebartels, C., van Montagu, M., Inze, D. and van Camp, W. (1997) Catalase is a sink for H_2O_2 and is indispensable for stress defence in C3 plants. *EMBO J.* **16**: 4806–4816.

Chapter 25

Production of transgenic tobacco plants with arginine decarboxylase under the control of an inducible promoter

Carles Masgrau, Mireia Panicot, Alexandra Cordeiro, Cristina Bortolotti, Oscar Ruiz, Teresa Altabella and Antonio Fernandez Tiburcio

1. Background

Polyamines are considered essential for growth and development (Tabor and Tabor, 1984). They are present in all living organisms, and their subcellular localization probably determines their function. The chemical definition of polyamines identifies them as aliphatic molecules positively charged at physiological pH, with a flexible three-dimensional structure that allows them to interact simultaneously with different negatively charged groups in macromolecules. This property differentiates these compounds from other cations like K^+, Ca^{2+} or Mg^{2+}.

Polyamines are present in cells as free bases interacting with macromolecules like DNA and RNA by charge affinity (Cohen, 1998), or covalently bound to many cellular compounds, like proteins (Del Duca et al., 1995; Folk, 1980), membrane components (Schuber, 1989; Singh et al., 1995), pectic substances in the cell wall (Messiaen et al., 1997). In plants, an important proportion is covalently bound to phenolic compounds derived from cinnamic acid (Bors, 1989; Martin-Tanguy, 1997).

The biosynthesis of the most common polyamines (putrescine, spermidine and spermine) is well known at the biochemical level, but new data about the genetic regulation and the exact localization of the biosynthetic enzymes are appearing with the use of recently produced molecular probes (Malmberg et al., 1998; Tiburcio et al., 1997). The diamine putrescine can be formed directly from ornithine by ornithine decarboxylase (ODC). Alternatively, in plants putrescine can also be indirectly synthesized from the decarboxylation of arginine by arginine decarboxylase (ADC), through two consecutive enzymatic steps involving the rapid metabolization of agmatine and N-carbamyl-putrescine.

With the addition of an aminopropyl moiety to putrescine, spermidine and spermine are formed sequentially by the enzymes spermidine synthase and spermine synthase.

Plant Responses to Environmental Stress, edited by M.F. Smallwood, C.M. Calvert and D.J. Bowles.
© 1999 BIOS Scientific Publishers, Oxford.

The donor of the aminopropyl group is the decarboxylated form of *S*-adenosyl-methion-ine (dcSAM), and its precuror (SAM) is a key substrate for ethylene biosynthesis (Adams and Yang, 1979). The flux of SAM to polyamine synthesis is controlled by the enzyme *S*-adenosyl-methionine decarboxylase and is proposed to be important in the relation between the two biosynthetic pathways (Apelbaum, 1981).

The first biochemical studies were performed with K^+-deficient oat plants (Smith, 1963). Later it was discovered that many cereals develop a rapid response to abiotic stress accumulating putrescine massively through induction of ADC activity (Slocum and Flores, 1991). Generally, ADC activity can be considered the main biosynthetic putrescine supply in morphogenesis, response to abiotic and biotic stresses, and during flowering initiation (Galston *et al.*, 1997). ODC activity proportionates polyamines dur-ing cell division, and is localized in meristematic tissues (Slocum and Flores, 1991).

Less is known about polyamine catabolism in plants. Diamino oxidase and polyamine oxidase have been described in dicots and monocots. These oxidases are implicated in the production of free radicals and toxic aldehydes and in lignification processes of cell wall (Angelini, 1993). In animals, there are studies reporting that catabolism of polyamines is through its acetyl derivatives, and that the genes involved are strongly regulated (Persson *et al.*, 1996). Recently, acetylated polyamines have been identified in plants (Caffaro *et al.*, 1993; De Agazio *et al.*, 1996).

During the last few years, with the cloning of the cDNAs coding for the polyamine biosynthetic enzymes, transgenic plants have been produced with different degrees of success. Homeostasis of polyamine levels in plants is probably tightly regulated. More work is needed to dissect the genomic structure and regulation of the genes, together with a deeper understanding of the biochemistry of the enzymes and their localization (Malmberg *et al.*, 1998). The use of inducible promoters like the TetR system (Gatz *et al.*, 1992) and/or tissue specific promoters to produce transgenic plants could facilitate the manipulation of polyamine contents in a controlled pattern of time and localization.

2. Molecular approach to the role of polyamine accumulation in abiotic stress

In order to investigate the relation between polyamine metabolism and abiotic stress we studied the effects of over-expressing ADC in plants. We have generated transgenic tobacco plants containing the oat ADC coding region (Bell and Malmberg, 1990), under the control of a tetracycline-inducible promoter (TripleX) (Gatz *et al.*, 1992). We have studied the relation between high levels of transgene expression and the appearance of toxicity due to putrescine accumulation. Plants from line 1.1 at T_0 and T_1 generation pre-sented high levels of transgene expression upon induction of the TripleX promoter, and displayed an altered phenotype that was accompanied by increases in ADC activity and high levels of putrescine and spermidine (Masgrau *et al.*, 1997). On the other hand, plants from line 52 did not present high levels of transgene-inducible expression, and no toxic phenotype appeared. However, upon induction, ADC activity and putrescine levels increased moderately.

With this in mind, we decided to study in more detail line 52 by analysing T_1 progeny in order to select homozygous plants with high levels of transgene expression. By northern analysis we selected two individuals: 52.4 (moderate expressor) and 52.6 (high expressor), and seeds were obtained from the auto-pollination of these two plants. By germination experiments in selective medium, Mendelian segregation of the T-DNA inserted in the genome of these lines was studied. Our results indicate that increases in ADC activity and putrescine levels, due to high transgene expression (line 52.6), led to toxicity effects similar to the ones observed in line 1.1.

With the availability of the tobacco cDNAs coding for the polyamine biosynthetic enzymes (Cordeiro, 1997), we also produced transgenic tobacco plants with the homologous ADC cDNA under the control of the Tet-inducible promoter, in sense and anti-sense orientation. Currently, we are analysing the T_1 generation of plants from selected lines that highly over-express the transgene in sense orientation. Preliminary results corroborate our previous data from transgenic tobacco plants with the oat ADC cDNA. We are also determining ADC activity and polyamine levels in transgenic plants with the tobacco ADC cDNA in antisense orientation. In some transgenic lines, induction of antisense transgene expression leads to a reduction of ADC activity and putrescine levels, and it is planned to study in detail the effects of antisense transgene on putrescine accumulation.

3. Production of tobacco transgenic plants with the oat arginine decarboxylase-coding region under the control of the inducible promoter TripleX

The putative transgenic plants were analysed by polymerase chain reaction (PCR) and southern analysis to confirm the presence of the oat ADC transgene in the tobacco genome. Inducible over-expression of the transgene was confirmed by northern analysis of total RNA from leaves of plants induced during 4 days with tetracycline applied in the feed solution. High and moderate expressing transgenic lines were selected for further analysis. These lines were used to establish the specificity and inducibility of the oat ADC transgene expression. For this, we performed an experiment incubating leaf discs in the presence or absence of tetracycline. The results obtained indicated that, upon tetracycline incubation, the oat ADC was completely induced after 48 h, and that without antibiotic, the expression was repressed. In the same leaf disc experiment we determined ADC activity, observing that ADC activity clearly increased in lines with higher transgene expression (Masgrau et al., 1997).

4. Analysis of T_1 transgenic plants from line 1.1

Line 1.1 was chosen for having high levels of transgene expression and for presenting a 16-fold increase in ADC activity under inductive conditions. After 3–5 weeks of tetracycline induction, polyamine content was analysed in leaf tissues. Comparing to non-induced controls, a general increase of polyamines was observed in the induced plants. Conjugated fractions (SH and PH) increased by two-fold, while free polyamines (S) remained unchanged. Moreover, polyamine changes were accompanied by toxicity symptoms, including interveinal chlorosis and necrosis of mature leaves. Abnormal development of young leaves followed by leaf necrosis together with a generalized collapsing of the mesophyll areas were observed in the most affected plants (Masgrau *et al.*, 1997).

5. Analysis of T_2 transgenic plants from line 52

Since T_1 induced plants from line 52 did not show any altered phenotype after 3–5 weeks of tetracycline treatment, and presented lower expression levels of the transgene compared to line 1.1, we decided to obtain segregants from this moderate expressor line. These segregants were analysed by northern analysis using total RNA from leaf discs. A probable homozygous plant, 52.6, was selected for its high level of transgene expression compared to plants with moderate inducible expression (e.g. 52.4).

Seeds from auto-pollinated plants 52.6 and 52.4 were germinated in selection media with antibiotics. As expected, 100% of seeds from line 52.6 germinated, while only 75% of seeds germinated for line 52.4. These results correspond to a Mendelian segregation of the antibiotic resistance trait contained in the T-DNA used for the production of the transgenic plants. Preliminary southern analysis determined the presence of one copy of the T-DNA inserted in the genome of the T_0 transgenic tobacco plant line 52. For plants of line 52.6, the exact number of T-DNA copies needs to be determined, to confirm the experiments performed with seed germination in selection media.

Progeny of the two plants (52.4 and 52.6) were grown to measure ADC activity upon induction of the transgene expression. The results showed that ADC activity increased 12-fold in induced plants from line 52.6, compared to an eight-fold increase in some plants of line 52.4. When the induction of the oat ADC in plants derived from line 52.6 was prolonged for 3–5 weeks, phenotype alterations appeared and changes in polyamine contents were similar to those observed in line 1.1. Total polyamine contents increased two-fold, mainly due to increases in putrescine and spermidine conjugated forms. The phenotypes observed ranged from generalized growth inhibition of aereal parts and roots to necrosis of mature leaves and abnormal development of young leaves.

6. Accumulation of endogenous putrescine (and spermidine) leads to toxicity symptoms

Results from analysis of lines 1.1 and 52.6 clearly indicate that toxic phenotype appears when the oat ADC transgene is highly over-expressed. Accumulation of putrescine and spermidine over the normal levels in a specific tissue and at a specific stage of plant development (e.g. young and mature leaves) seems to be the cause. Even though in these lines ADC activity increased more than 10-fold when the phenotype appeared, total putrescine increased only by two- to three-fold. To check the possibility that by over-expressing ADC we were provoking an arginine deficiency in the induced plants, we supplemented the feed solution with 1 mM arginine. Upon induction of the transgene, the toxicity phenotype developed earlier in plants from line 52.6 showing chlorosis symptoms and growth inhibition. In these plants, ADC activity increased by 30-fold compared to the 10-fold increase measured in control induced plants, grown in normal feed solution.

Putrescine catabolic processes might lead to an accumulation of toxic aldehydes and free radicals (DiTomaso, 1989) due to diamino oxidase activity. The use of amino-guanidine, an inhibitor of diamino oxidase, had experimental drawbacks because it led to strong growth inhibition of control plants, so that we could not obtain conclusive results. However, transgenic plants (line 52.6) induced in the presence of amino-guanidine presented rapidly strong chlorosis and necrosis dying after 3 weeks of treatment.

Data suggest that endogenous putrescine accumulation is highly toxic to plants. At the moment experiments are being conducted to determine DAO activity in tobacco plants in order to establish the role played by putrescine catabolism on the observed phenotypes.

7. Production of tobacco transgenic plants with the homologous tobacco arginine decarboxylase cDNA, in sense and antisense orientation, under the control of the inducible promoter TripleX

Transgenic tobacco plants were produced with both sense (As) and antisense (Aa) orientation of the homologous ADC cDNA. Regenerants were submitted to PCR screening to confirm the presence of the T-DNA and for each construct 20–25 transgenic lines were obtained. Induction of the transgene expression has been performed treating leaf discs of the transgenic As and Aa plants with tetracycline. Northern analysis of total RNA from induced leaf discs allowed the selection of transgenic As lines with higher expression levels of the transgene. Selected plants were induced to measure ADC activity increases (*Table 1*) and polyamine levels. Preliminary results indicate that the over-expression of the transgene and the resulting alterations in polyamine levels correlate with the observed phenotypic alterations (*Table 1*).

The same procedure was used for Aa transgenic plants. When we compared endogenous ADC mRNA levels and antisense transgene expression, three different transcriptional patterns were found. In *Table 1* results about the effects of the antisense transgene on ADC activity are presented.

Our results show that the over-expression of ADC in tobacco plants leads to an increase in RNA levels, ADC activity and putrescine levels, and consequently a toxic phenotype is observed, characterized by chlorosis and necrosis in mature leaves.

These effects are similar to those observed previously in our laboratory, when oat leaves were incubated under osmotic stress conditions. Similar results were also obtained many years ago by Richards and Coleman (1952) in barley plants growing under K^+ deficiency. In both cases, accumulation of putrescine as a result of an increase in ADC activity was observed.

In addition, in induced tobacco transgenic plants with the oat ADC, ultrastructural studies evidenced plasmolysis in mesophyll cells. This effect is similar to that observed under osmotic and K^+ deficiency stresses.

Table 1. Preliminary results from T$_1$ transgenic plants with the homologous arginine decarboxylase (ADC) cDNA in sense and antisense orientation

Line	Northern	ADC activity (%)	Phenotype
ADCs (As)			
13	+ +	+40	Growth inhibition
21	+	–	Growth inhibition, leaf necrosis
ADCa (Aa)			
5	s+	–55	No phenotype
	a+ +		
13.1	s+	=	No phenotype
	a–		
16	s–	–	Growth inhibition,
	a–		twisted leaves

Plants grown in hydroponic conditions were induced with tetracycline for 3–4 weeks. For northern analysis expression, levels are indicated as: ++, high expression; +, normal expression; –, reduced expression; a, antisense; s, sense.

We are currently performing experiments with antisense ADC transgenic plants in order to determine whether plants with reduced ADC activity and reduced putrescine contents have an altered response to abiotic stresses like K^+ deficiency, osmotic and/or salinity stress.

References

Adams, D.O. and Yang, S.F. (1979) Ethylene biosynthesis: identification of 1-aminocyclo-propane-1-carboxylic acid as an intermediate in the conversion of methionine to ethylene. *Proc. Natl. Acad. Sci. USA* **76:** 170–174.

Angelini, R., Bragaloni, M., Federico, R., Infantino, A. and Porta-Puglia, A. (1993) Involvement of polyamines, diamine oxidase and peroxidase in resistance of chickpea to *Ascochyta rabiei. J. Plant Physiol.* **142:** 704–709.

Apelbaum, A., Burgoon, A.C., Anderson, J.D., Lieberman, M., Ben-Arie, R. and Mattoo, A.K. (1981) Polyamines inhibit biosynthesis of ethylene in higher plant tissue and fruit protoplasts. *Plant Physiol.* **68:** 453–456.

Bell, E. and Malmberg, R.L. (1990) Analysis of a cDNA encoding arginine decarboxylase from oats reveals similarity to *Escherichia coli* arginine decarboxylase and evidence of protein processing. *Mol. Gen. Genet.* **224:** 431–436.

Bors, W., Langerbartels, C., Michel, C. and Sanderman, H. (1989) Polyamines as radical scavengers and protectants against ozone damage. *Phytochemistry* **28:** 1589–1595.

Caffaro, S., Scaramagli, S., Antognoni, F. and Bagni, N. (1993) Polyamine content and translocation in soybean plants. *J. Plant Physiol.* **141**: 563–568.

Cohen, S.S. (1998) *A Guide to Polyamines.* Oxford University Press, New York.

Coleman, T.G. and Richards, F.J. (1956) Physiological studies in plant nutrition. XVIII. Some aspects of nitrogen metabolism in barley and other plants in relation to potassium deficiency. *Ann. Bot.* **20:** 393–409.

Cordeiro, A. (1997) Molecular and biochemical analysis of methylglyoxal-bis(guanylhydrazone) resistance and of polyamine metabolism in relation to cell division in *Nicotiana tabacum.* PhD thesis, University of Cologne.

De Agazio, M., Grego, S., Zacchini, M., De Cesare, F., Cellai, L., Rizea-Savu, S. and Silvestro, L. (1996) 1–*N*–acetylspermidine in roots of maize seedlings. *Science* 121:143–149.

Del Duca, S., Beninati, S. and Serafini-Fracasini, D. (1995) Polyamines in chloroplasts: identification of their glutamyl and acetyl derivatives. *Biochem. J.* **305:** 233–237.

DiTomaso, J.M., Shaff, J.E. and Kochien, L.V. (1989) Putrescine-induced wounding and its effects on membrane integrity and ion transport processes in roots of intact corn seedlings. *Plant Physiol.* **90:** 988–995.

Folk, J.E. (1980) Transglutaminases. *Annu. Rev. Biochem.* **49:** 517–531.

Galston, A.W., Kaur-Sawhney, R., Altabella, T. and Tiburcio, A.F. (1997) Plant polyamines in reproductive activity and response to abiotic stress. *Bot. Acta* **10:** 197–207.

Gatz, C., Frohgberg, C. and Wendenburg, R. (1992) Stringent repression and homogeneous de-repression by tetracycline of a modified CaMV 35S promoter in intact transgenic tobacco plants. *Plant J.* **2:** 397–404.

Malmberg, R.L., Watson, M.B., Galloway, G.L. and Yu, W. (1998) Molecular genetic analyses of plant polyamines. *Crit. Rev. Plant Sci.* **17:** 199–224.

Martin-Tanguy, J. (1997) Conjugated polyamines and reproductive development: biochemical, molecular and physiological approaches. *Physiol. Plant.* **100:** 675–688.

Masgrau, C., Altabella, T., Farràs, R., Flores, D., Thompson, A.J., Besford, R. and Tiburcio, A. (1997) Inducible overexpression of oat arginine decarboxylase in transgenic tobacco plants. *Plant J.* **11:** 465–473.

Messiaen, J., Cambier, P. and Cutsem, P.V. (1997) Polyamines and pectins. I. Ion exchange and selectivity. *Plant. Physiol.* **113:** 387–395.

Persson, L., Svensson, F. and Lövkvist, W.-E. (1996) Regulation of polyamine metabolism. In: *Polyamines in Cancer: Basic Mechanisms and Clinical Approaches* (ed. K. Nishioka). RG Landes Co., Austin, TX, pp. 14–43.

Richards, F.J. and Coleman, R.G. (1952) Occurrence of putrescine in K-deficient barley. *Nature* **170:** 460.

Schuber, F. (1989) Influence of polyamines on membrane functions. *Biochem. J.* **260:** 1–10.

Singh, S.S., Chauhan, A., Brockerhoff, H. and Chauhan, V.P.S. (1995) Differential effects of spermine on phosphatidyl-inositol 3-kinase and phosphatidyl-inositol phosphate 5-kinase. *Life Sci.* **57:** 685–694.

Slocum, R.D. and Flores, H.E. (eds) (1991) *Biochemistry and Physiology of Polyamines in Plants.* CRC Press, Boca Raton, FL.

Smith, T.A. (1963) L-Arginine carboxy-lyase of higher plants and its relation to potassium nutrition. *Phytochemistry.* **2:** 241–252.

Tabor, C.W. and Tabor, H. (1984) Polyamines. *Annu. Rev. Biochem.* **53:** 749–790.

Tiburcio, A.F., Altabella, T., Borrell, A. and Masgrau, C. (1997) Polyamine metabolism and its regulation. *Physiol. Plant.* **100:** 664–674.

Chapter 26

The peel of the *Citrus* fruit: a good source for stress-related genes/proteins

Maria T. Sanchez-Ballesta, Xavier Cubells, José M. Alonso, Maria T. Lafuente, Lorenzo Zacarias and Antonio Granell

1. Background

Fruits are important for frugivorous animals as food but also for plants as they provide protection to the seed normally contained within and a vehicle for its dissemination. The peel of the fruit constitutes the organ interface with the environment. It protects the fruit from abiotic stresses such as irradiation, drying, *et cetera*, and also from biotic stresses namely pathogens, while it must attract herbivores to enable seed dispersal. *Citrus* fruits are defined as herperidiums (a berry) with a fragrant rind. We are interested in the changes occurring in the flavedo (outer coloured layer of the fruit) during fruit maturation and in the study of the molecular changes occurring in this part of the fruit leading to protection against environmental stresses and to attraction by herbivores.

Within the pericarp of the citrus fruit, the exocarp or outer rind is composed of different tissues ontogenically related to the leaf. A number of characteristics of these tissues indicates the role of this part of the fruit in protecting the fruit and making it appealing. A thick cuticle covers the surface and intrudes between the walls of external layers of cells. Crystal idioblasts and oil glands occur within the tissue under the epidermis (Schneider, 1968). During ripening, chloroplasts of the parenchyma cells are converted to chromoplasts giving the fruits their characteristic colour. This is due mainly to degradation of chlorophyll and accumulation of carotenoids (Goldschmidt, 1988).

The flavedo of the citrus fruit synthesizes and accumulates abundant antioxidant small molecules including L-ascorbic acid (vitamin C) and carotenoids. Interestingly, while *Citrus* fruits are one of the best sources for ascorbate, only one-quarter of it is found in the juice and three-quarters is in the peel, mainly accumulated in the flavedo (Ting and Attaway, 1971).

The importance of ascorbate for protection against environmental stresses has been reinforced with the increased susceptibility to environmental stresses shown by *Arabidopsis* mutants deficient in ascorbate (Conklin *et al.*, 1996). Flavedo is also rich

Plant Responses to Environmental Stress, edited by M.F. Smallwood, C.M. Calvert and D.J. Bowles.
© 1999 BIOS Scientific Publishers, Oxford.

in phenolic compounds and flavonoids that may act as sunscreens. These compounds, as demonstrated recently in *Arabidopsis*, protect the plant against UV light and oxidative stress (Landry *et al.*, 1995; Rao and Ormrod, 1995). It is also worth mentioning that a direct correlation exists between the levels of ascorbic acid in the flavedo and the resistance of this tissue against different pathogens. During fruit storage the levels of ascorbic acid [and also of the phytoalexin scoparone (Ben-Yehoshu *et al.*, 1992; Kim *et al.*, 1991)] decrease while the susceptibility to pathogens increases.

2. Genes expressed in the flavedo during maturation and in response to environmental stresses

The isolation of the gene expressed in the flavedo during maturation has been conducted in our laboratory by using differential screening and differential display techniques. Based on the sequence and expression analysis we highlight the existence of three broad classes of genes expressed in the flavedo.

2.1 Genes involved in the transformation of the chloroplast into chromoplast

Genes corresponding to this class include the small subunit of Rubisco and chlorophyll *a/b* binding proteins whose transcript levels decrease during the transformation of the chloroplast to chromoplast. A chlorophyllase activity that is regulated by ethylene seems to be implicated in this process. Interestingly, the transcript for a non-photosynthetic ferredoxin (Alonso *et al.*, 1995b) also accumulates in the flavedo where it appears to be required for chlorophyll catabolism occurring during de-greening (Cubells *et al.*, 1998).

2.2 Genes involved in other aspects of the ripening process

An increase in the sucrose phosphate transcript levels in the flavedo similar to that reported in juice sacs (Komatsu *et al.*, 1996) was also detected in our laboratory (A. Granell, unpublished data). This suggests that sucrose accumulation occurring in the flavedo may be regulated at the transcriptional level of enzymes involved in its biosynthesis.

2.3 Constitutive and inducible stress-related genes

In the flavedo, some typical stress-inducible genes (like late embryogenesis abundant proteins, LEA and non-specific lipid transfer protein, LTP, etc.) are expressed at high levels (M.T. Sanchez-Ballesta, T. Lafuente, L. Zacarius and A. Granell, unpublished data). Does this reflect the fact that the flavedo is a stressed tissue under normal growth conditions, or were fruits afflicted by undetected stresses? To answer this question we analysed the transcript levels in leaves. The levels of most of these transcripts in leaves were very low or undetectable despite the fact that leaves can respond to externally applied stress by accumulating these mRNAs. Furthermore, the flavedo still remains sensitive to environmental stresses deduced from the increase in the levels of a number

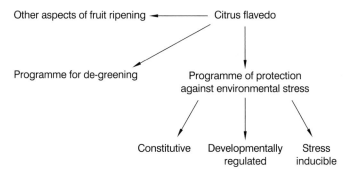

Figure 1. Programmes co-existing during flavedo maturation in citrus fruit.

of stress-related mRNAs which can be obtained after cold stress. These results may reflect that a constitutive versus an adaptive response to different stress is operating in each organ or that different organs contain different basal levels of stress-related genes.

Stresses such as cold can induce the accumulation of specific mRNAs in the flavedo which are absent or present at very low levels in unstressed fruits. These mRNAs were identified by differential screening techniques. The sequence for some of these mRNAs is not very informative but some other mRNAs of this group show similarities to glucanases, phytoalexin biosynthetic enzymes, *et cetera*. This indicates that in the flavedo, pathogen and cold (and also other stresses) response pathways may be cross-talking to some extent. In support of this idea, LEA (Cai *et al.*, 1995), and LTPs (Molina *et al.*, 1996; Treviño and O'Connell, 1998) are known to be regulated by environmental stresses (as in our case) and have a role against pathogens (Molina *et al.*, 1993), although other roles are also possible such as the one for LTP in wax deposition (Pye and Kolattukudy, 1995).

The observation that some of the genes regulated during flavedo maturation can be regulated in the same manner by exogenous ethylene indicates that this hormone may participate in the control of some aspects of flavedo maturation and more specifically those involved in flavedo de-greening and at least in some of the responses to environmental stress (Alonso *et al.*, 1995a, 1995b; Cubells *et al.*, 1998). Interestingly, the mRNAs for a vacuolar processing enzyme also accumulates to high levels in the flavedo (Alonso and Granell, 1995). This vacuolar processing protease may be required for the processing of vacuolar targeted activities such as glucanases, quitinases *et cetera* that are known to accumulate in biotic and abiotically stressed plants (including *Citrus*).

We are currently isolating genomic sequences for the genes that show high expression in the flavedo. The promoter sequences of these genes could be useful to direct specific expression in the flavedo of the protein of interest in transgenic *Citrus* plants. *Citrus* transformation has been developed recently and it is now possible to introduce genes of interest into this important crop (Peña *et al.*, 1995). Over-expression of *Citrus* stress-inducible genes under the control of a strong flavedo constitutive promoter is one of our first targets. Also we plan to perform heterologous transformations and the isolation of homologous genes in *Arabidopsis*. A reverse genetic approach in *Arabidopsis* for a number of *Citrus* stress-related mRNAs that show no sequence homology to known proteins could indicate their function.

Acknowledgements

We are grateful to the Spanish Ministry of Science and Education and CICYT for financial support for this work.

References

Alonso, J.M. and Granell, A. (1995). A putative vacuolar processing protease is regulated by ethylene and also during fruit ripening in *Citrus* fruit. *Plant Physiol.* **109:** 541–547.

Alonso, J.M., Chamarro, J. and Granell, A. (1995a) Evidence for the involvement of ethylene in the expression of specific RNAs during maturation of the orange, a non climacteric fruit. *Plant Mol. Biol.* **29:** 385–390.

Alonso, J.M., Chamarro, J. and Granell, A. (1995b) A non-photosynthetic gene is induced by ethylene in *Citrus* organs. *Plant Mol. Biol.* **29:** 1211–1221.

Ben-Yehoshua, S., Rodoy, V., Kim, J.J. and Carmeli, S. (1992) Preformed and induced antifungal materials of citrus fruits in relation to the enhancement of decay resistance by heat and ultraviolet treatments. *J. Agr. Food. Chem.* **40:** 1217–1221.

Cai, Q., Moore, G.A. and Guy, C.L. (1995) An unusual group 2 LEA family in citrus responsive to low temperature. *Plant Mol. Biol.* **29:** 11–23.

Conklin, P.L., Williams, E.H. and Last, R.L. (1996) Environmental stress sensitivity of an ascorbic acid-deficient *Arabidopsis* mutant. *Proc. Natl Acad. Sci. USA* **93:** 9970–9974.

Cubells, X., Sanchez-Ballesta, M.T., Alonso, J.M. and Granell, A. (1999) Ethylene perception and response in *Citrus* fruit. In: *Biology and Biotechnology of the Plant Hormone Ethylene II* (eds A.K. Kanellis *et al*). Kluwer Academic Press, Dordrecht, The Netherlands (in press).

Goldschmidt, E.E. (1988) Regulatory aspects of chloro-chromoplast interconversion in senescing citrus fruit peel. *Isr. J. Bot.* **37:** 123–130.

Kim, J.J., Ben-Yehoshua, S., Shapiro, B., Henis, Y. and Carmeli, S. (1991) Accumulation of Scoparone in heat-treated lemon fruit inoculated with *Penicillium digitatum* Sac. *Plant. Physiol.* **97:** 880–885.

Komatsu, A., Takanokura, Y., Omura, M. and Akihama, T. (1996) Cloning and molecular analysis of cDNAs encoding three sucrose phosphate synthase isoforms from a citrus fruit (*Citrus unshiu* Marc.). *Mol. Gen. Genet.* **151:** 346–351.

Landry, L.G., Chapple, C.C.S. and Last, R.L. (1995) Arabidopsis mutants lacking phenolic sunscreens exhibit enhanced ultraviolet-B injury and oxidative damage. *Plant Physiol.* **109:** 1159–1166.

Molina, A., Diaz, I., Vasil, I.K., Carbonero, P. and Garcia-Olmedo, F. (1996) Two cold-inducible genes encoding lipid transfer protein LTP4 from barley show differential responses to bacterial pathogens. *Mol. Gen. Genet.* **252:** 162–168.

Molina, A., Segura, A. and García-Olmedo, F. (1993) Lipid transfer proteins (nsLTPs) from barley and maize leaves are potent inhibitors of bacterial and fungal plant pathogens. *FEBS Lett.* **316:** 119–122.

Peña, L., Cervera, M., Juarez, J., Navarro, A., Pina, J.A., Duran-Vila, N. and Navarro, L .(1995) *Agrobacterium*-mediated transformation of sweet orange and regeneration of transgenic plants. *Plant Cell Rep.* **14:** 616–619.

Pye, J. and Kolattukudy, P.E. (1995) The gene for the major cuticular wax-associated protein and three homologous genes from broccoli (*Brassica oleracea*) and their expression patterns. *Plant J.* **7:** 49–59.

Rao, M.V. and Ormrod, D.P. (1995) Impact of UVB and O$_3$ on the oxygen free radical scavenging system in *Arabidopsis thaliana* differing in flavonoid biosynthesis. *Photochem. Photobiol.* **64:** 719–726.

Schneider, H. (1968) The anatomy of *Citrus*. In: *The Citrus Industry,* Vol. II (eds W. Reuther, L.D. Batchelor and H.J. Webber). University of California Centennial Public, pp. 1–23.

Ting, S.V. and Attaway, J.A. (1971) Citrus fruits. In: *The Biochemistry of Fruit and their Products* (ed. A.C. Hulme). Academic Press, London, pp. 107–171.

Treviño, M.B. and O'Connell, M.A. (1998) Three drought-responsive members of the nonspecific lipid-transfer protein gene family in *Lycopersicon pennellii* show different developmental patterns of expression. *Plant Physiol.* **117:** 1461–1468.

Chapter 27

An abiotic stress response in cassava: post-harvest physiological deterioration

John R. Beeching, Holger Buschmann, Rocío Gómez-Vásquez, Yuanhuai Han, Carlos Iglesias, Hongying Li, Kim Reilly and Maria Ximena Rodriguez*

1. Background

Cassava (*Manihot esculenta* Crantz), a member of the Euphorbiaceae which includes castor bean (*Ricinus communis* L.) and the rubber tree (*Hevea brasiliensis* L.), is the most important tropical root crop. It is a perennial shrub of 1–5 m tall and is grown for its starchy roots. It was probably domesticated in the Amazon basin about 7000 years ago (Lathrap, 1970), since when its cultivation has spread through the tropical regions of America, and subsequent to its colonization, to tropical regions of Africa and Asia (Cock, 1985).

As a crop, cassava is propagated via stem cuttings and the swollen roots are generally harvested after 6–12 months, although they can be left in the soil until needed. Cassava will grow on impoverished soil and on soils that contain high concentrations of toxic compounds, such as aluminium and magnesium, which would inhibit the growth of other crops such as beans or maize. Cassava is a hardy and robust crop, and its cultivation requires minimal technological input (Cock, 1985). These factors account for cassava's important role as a subsistence crop for the poor, especially in Africa, and for its critical importance in the event of famine or war. Increasingly, especially in South-East Asia, cassava is being grown commercially for starch or for the production of chips for export as animal feed. Cassava is the staple food of over 500 million people; in some African countries, such as the Democratic Republic of Congo, cassava can supply over 50% of the dietary carbohydrate (CIAT, 1992). In 1991, world production of cassava was 162 million tonnes, of which 75 million was in Africa (Wenham, 1995). With populations expected to rise in tropical regions it is anticipated that production of cassava will increase.

2. Constraints on production and use of cassava

The international Cassava Biotechnology Network (CBN) has identified major

*All authors contributed equally to this chapter.

Plant Responses to Environmental Stress, edited by M.F. Smallwood, C.M. Calvert and D.J. Bowles.
© 1999 BIOS Scientific Publishers, Oxford.

205

constraints to the increased production and utilization of cassava (Puonti-Kaerlas, 1998). These constraints include the biotic stresses imposed by the major cassava pathogens: African Cassava Mosaic Virus (ACMV) and bacterial blight (*Xanthomonas axonopodis*), and the abiotic stress response of post-harvest physiological deterioration (PPD). Within 24–72 h after harvesting the roots start to deteriorate rendering them unpalatable and unmarketable; this is an abiotic response and is not associated with any microorganisms (Rickard and Coursey, 1981; Wheatley and Gomez, 1985). Cassava, therefore, must be consumed or processed immediately upon harvesting. In a traditional rural setting PPD is not a significant problem; the roots can always be left in the ground until needed. However, with urban migration, unreliable transport and the need for uniform input material to processing plants, PPD has become a major constraint to the development of cassava production and marketing. While there do exist post-harvest treatments which can inhibit PPD, these are unpractical or uneconomical in the context within which cassava is produced and consumed (Beeching *et al.*, 1994).

The control of PPD is therefore a priority. CBN has suggested a shelf-life of up to 2 weeks as being a realistic goal to be achieved within the next decade. Although PPD has been the subject of research for a considerable period, the problem is far from understood. Therefore, there is an urgent need firstly to understand the nature of PPD from biochemical, molecular and genetic perspectives, and secondly, to attempt to control the phenomenon either via conventional breeding techniques or by genetic modification. Research at Bath, in collaboration with CIAT (Centro Internacional de Agricultura Tropical) in Colombia, is currently directed towards understanding the problem and to identifying potential key points for its control (Beeching *et al.*, 1998).

3. Aspects of post-harvest physiological deterioration

Harvesting necessarily involves the detachment of the root from the plant, other small wounds and abrasions inevitably accompany the harvesting process. It is believed that it is these small wound sites which provide the focal points for the initiation of PPD (Booth, 1976). The first visual symptom of PPD is a blue-black discoloration of the vascular tissue within the root; for this reason cassava is often broken open in the market to reveal its internal quality. However, prior to the appearance of this discoloration there is a strong fluorescence of the root under UV light. Microscopically, coloured occlusions and tyloses formed from the xylem parenchyma are observed; these block adjacent xylem vessels (Rickard and Gahan, 1983).

PPD is accompanied by an increase in respiration; there is also some mobilization of starch into sugars, and acid invertase activity increases. Production of the phytohormone ethylene increases; ethylene is held to play a co-ordinating role in senescence, wounding and ripening events in other plants (Ecker and Davis, 1987). The activities of other enzymes such as catalase, dehydrogenases, peroxidases, phenylalanine ammonia-lyase (PAL) and polyphenol oxidase also increase during PPD (Hirose, 1986; Plumbley *et al.*, 1981; Uritani *et al.*, 1984). These data also suggest that PPD is an active process involving changes in gene expression. Certainly, the application of cyclohexamide, an inhibitor of protein synthesis, inhibits the changes associated with PPD (Uritani *et al.*,

1984); and we have shown, using *in vivo* labelling of proteins, that there is a massive *de novo* increase in protein synthesis, including novel proteins, during the first 48 h of PPD (Beeching *et al.*, 1995).

Low molecular mass compounds are also produced or change in abundance during the time course of PPD. Phenolic and polyphenolic compounds, including the coumarin scopoletin, increase in the roots after harvesting (Uritani *et al.*, 1983). Scopoletin fluoresces bright blue under UV light and probably contributes significantly to the fluorescence observed in the early stages of deterioration in the roots (Wheatley and Schwabe, 1985). This compound has been shown to have anti-microbial activities in *Hevea brasiliensis* (Giesemann *et al.*, 1986) and may play a similar role in cassava. Steroids (probably oxidation products of original membrane sterols) and diterpenes are other stress metabolites that have been detected (Sakai and Nakagawa, 1988). During PPD there is a decline in total phospholipids and glyceroglycolipids, and a concomitant increase in sterol-containing lipids (Lalaguna and Agudo, 1989). These changes could be a result of the breakdown of membrane lipids in the affected tissues. The oxidation of membrane lipids in damaged tissues can lead to the formation of long chain fatty acids including linolenic, a precursor in the biosynthesis of jasmonic acid, which has been shown to function as a signalling molecule in wound responses in other plants (Farmer and Ryan, 1992).

4. Commonality of post-harvest physiological deterioration and biotic/abiotic stress responses

From the above we can conclude that PPD in cassava is an active process involving changes in gene expression, and that it includes processes common to wound responses in other plants. The initial damage to a plant by wounding releases cell wall fragments and membrane components that can act as initial local signals. These in turn induce the production of other signals, such as jasmonates, salicylic acid, hydrogen peroxide, the peptide systemin, electrical and hydraulic signals, which act locally to the wound or systemically throughout the plant. The net effects of these signals are attempts to defend and repair the wound site and to prepare the plant for the potential extension of the wounding process. A wound is a potential site for invasion by microorganisms; therefore, components of the wound response are the production of anti-microbial enzymes, such as chitinase and β-1,3-glucanase and low molecular mass compounds, such as flavonoids and phenolics, some of which have anti-microbial activity. In addition, barriers are formed at the wound site to prevent its further extension and to heal the plant; the formation of lignin, suberin and hydroxyproline-rich glycoproteins (HRGPs) play major roles in this barrier formation process. General phenylpropanoid metabolism, of which PAL is the key entry enzyme, plays pleiotropic roles in wound responses; outputs from phenylpropanoid metabolism include anti-microbial phenolic compounds, the signalling molecule salicylic acid, and barrier-forming lignin and suberin. As we have seen above, many of these wound response components are present in cassava PPD, except that wound healing is inadequate. It is possible that this inadequate sealing of the wound sites leads to the continual production of the initial wound-induced signals, thereby precipitating continuous cascades of wound responses throughout the root. These responses being observed as PPD.

5. Molecular approaches to post-harvest physiological deterioration prevention

In order to more fully understand the process of PPD and to approach the identification of potential control points, via either breeding or genetic modification, we have initiated a multi-disciplinary research programme studying the biochemistry, molecular biology and genetics of the phenomenon.

A potentially productive way into the problem of PPD is to clone genes from cassava, which are known to be wound-related in other plants. To this end a cDNA library was constructed in λgt10 using mRNA isolated from cassava roots 48 h after harvesting. Using heterologous probes we isolated a range of clones from this library. These cDNAs were subsequently sub-cloned into plasmids and sequenced. The clones include two distinct PALs, and one each of HRGP, β-1,3-glucanase, 1-aminocyclopropane-1-carboxylate (ACC) oxidase (a key enzyme in ethylene biosynthesis), catalase and peroxidase. In addition, by sequencing anonymous clones from the library we have isolated a putative serine/threonine protein kinase. Southern blots using the PAL clones, and gene-specific parts of those clones, suggest that there are probably only two PAL genes present in the cassava genome. Though we have so far only cloned a partial cDNA of the protein kinase, this clone is of particular interest as it shows very high similarity to a receptor protein kinase from *Brassica*. The sequence of a full-length clone should reveal the nature of its receptor domain. Collaborators at CIAT in Colombia have included these clones in the genomic map of cassava, and are evaluating their use as markers in crosses between cassava cultivars which show high and low PPD sensitivity.

Screening a cassava genomic library (generously provided by Professor Monica Hughes, University of Newcastle) has led to the isolation of a genomic clone for ACC oxidase and clones for the two PAL genes. The two genes show high similarity to other plant PALs and the conservation of the typical intron position. One of these clones, PAL2, possesses an extensive promoter region, over 2 kb of which have been sequenced. We have made plasmid constructs using regions of the PAL2 promoter to drive the β-glucuronidase reporter gene. These have been bombarded into cassava using biolistics and shown to be transiently expressed in a variety of cassava tissues. We are currently embarking on the production of transgenic cassava plants, which should allow us to study the activity of the promoter for this important gene during PPD and other stress responses.

While PPD is a difficult character to score confidently as it is affected by environmental variables, we have available at Bath a range of cassava varieties that show high, medium and low PPD responses. Using high performance liquid chromatography (HPLC) and thin-layer chromatography (TLC) we have fractionated extracts from these varieties over a time course of PPD. This work has revealed varietal differences in the timing of appearance and abundance of a range of compounds. HPLC peaks of interest are currently being identified, particularly one that shows anti-microbial activity and a second whose presence is correlated with the onset of PPD.

Cell suspension cultures and elicitors have proved a powerful tool to dissect the potential range of a plant's stress responses. Using microbial cell wall extracts and cassava cell suspensions we have shown that cassava is capable of producing H_2O_2 in an oxidative burst. In addition, typical bi-phasic bursts are induced by incompatible pathogens. H_2O_2 has also been detected in the cassava root itself during PPD.

These results confirm that the major components of a normal wound response are present in cassava PPD. H_2O_2 and the ethylene produced via ACC oxidase are involved in signalling. The serine/threonine protein kinase could be part of a signal-transduction pathway. β-1,3-Glucanase and the low molecular mass compound identified by HPLC are anti-microbial. HRGPs are components of wound repair. PAL, catalase and peroxidase are involved in many aspects of wound responses. However, the wound repair aspects are insufficient to restore the integrity of the root. It is possible that these are not present either because they are not required in an organ which is not a propagule and has no function when detached from the plant, or because they have been accidentally bred out of the plant.

Acknowledgements

The research is supported by grants from the UK's Department for International Development. R.G.V. would like to thank Colciencias and CIF for support; Y.H. and H.L., the Department of Biology & Biochemistry; and M.X.R., Colfuturo.

References

Beeching, J.R., Dodge, A.D., Moore, K.G., Phillips, H.M. and Wenham, J.E. (1994) Physiological deterioration in cassava: possibilities for control. *Tropical Sci.* **34:** 335–343.

Beeching, J.R., Dodge, A.D., Moore, K.G. and Wenham, J.E. (1995) Physiological deterioration in cassava: an incomplete wound response? In: *The Cassava Biotechnology Network: Proceedings of the Second International Scientific Meeting*, 22–26 August 1994, Bogor, Indonesia (eds A.-M. Thro and W. Roca). CIAT, Cali, Colombia, pp. 729–736.

Beeching, J.R., Han, Y., Gomez-Vasquez, G., Day, R.C. and Cooper, R.M. (1998) Wound and defense responses in cassava as related to post-harvest physiological deterioration. *Recent Adv. Phytochem.* **32:** 231–248.

Booth, R.H. (1976) Storage of fresh cassava (*Manihot esculenta*). I. Post-harvest deterioration and its control. *Exp. Agric.* **12:** 103–111.

CIAT (1992) *Cassava Programme 1987–1991. Working Document 116.* CIAT, Cali, Colombia.

Cock, J.H. (1985) *Cassava: New Potential for a Neglected Crop.* Westfield Press, Boulder, CO.

Ecker, J.R. and Davis, R.W. (1987) Plant defence genes are regulated by ethylene. *Proc. Natl Acad. Sci. USA* **84:** 5202–5206.

Farmer, E.E. and Ryan, C.A. (1992) Octadecanoid precursors of jasmonic acid activate the synthesis of wound-inducible proteinase inhibitors. *Plant Cell* **4:** 129–134.

Giesemann, A., Biehl, B. and Lieberei, R. (1986) Identification of scopoletin as a phytoalexin of the rubber tree *Hevea brasiliensis. J. Phytopathol.* **117:** 373–376.

Hirose, S. (1986) Physiological studies on postharvest deterioration of cassava plants. *Japan Agric. Res. Q.* **19:** 241–252.

Lalaguna, F. and Agudo, M. (1989) Relationship between changes in lipid with ageing of cassava roots and senescence parameters. *Phytochemistry* **28:** 2059–2062.

Lathrap, D.W (1970) *The Upper Amazon.* Thames and Hudson, London.

Plumbley, R.A., Hughes, P.A. and Marriot, J. (1981) Studies on peroxidases and vascular discoloration in cassava root tissues. *J. Sci. Food Agric.* **32:** 723–731.

Puonti-Kaerlas, J. (1998) Cassava biotechnology. *Biotechnol. Genet. Engng Rev.* **15:** 329–364.

Rickard, J.E. and Coursey, D.G. (1981) Cassava storage. Part 1: storage of fresh cassava roots. *Trop. Sci.* **23:** 1–32.

Rickard, J.E. and Gahan, P.B. (1983) The development of occlusions in cassava (*Manihot esculenta* Crantz) root xylem vessels. *Ann. Bot.* **52:** 811–821.

Sakai, T. and Nakagawa, Y. (1988) Diterpenic stress metabolites from cassava roots. *Phytochemistry* **27:** 3769–3779.

Uritani, I., Data, E.S., Villegas, R.J., Flores, P. and Hirose, S. (1983) Relationship between secondary metabolism changes in cassava root tissue and physiological deterioration. *Agric. Biol. Chem.* **47:** 1591–1598.

Uritani, I., Data, E.S. and Tanaka, Y. (1984) Biochemistry of postharvest deterioration of cassava and sweet potato roots. In: *Tropical Root Crops: Postharvest Physiology and Processing* (eds I. Uritani and E.D. Reyes). Tokyo, JSSP, pp. 61–75.

Wenham, J.E. (1995) *Post-harvest Deterioration of Cassava. A Biotechnological Perspective.* FAO, Rome.

Wheatley, C. and Gomez, G. (1985) Evaluation of some quality characteristics in cassava storage roots. *Qual. Plant Foods Hum. Nutr.* **35:** 121–129.

Wheatley, C.C. and Schwabe, W.W. (1985) Scopoletin involvement in post-harvest physiological deterioration of cassava root (*Manihot esculenta* Crantz). *J. Exp. Bot.* **36:** 783–791.

Chapter 28

Cassava leaves: approach for studies on the applicability of proteins from leaves of cassava as an additive for food fortification or as a molecular sensor of abiotic stress

Marilia Penteado Stephan, Flavia Higino de Lima and Alessandro dos Santos Frazão

1. Background

The participation of leaves in the maintenance of the global alimentary chain is repre-sented by two processes of transforming inorganic to organic compounds in nature. Initially, the incorporation of atmospheric inorganic CO_2 into the organic machinery of living cells is a reaction that has the enzyme ribulose diphosphate carboxylase (RuBP carboxylase) as its main component. It represents one of the most important biochemi-cal reactions participating in the renewal of energy in the environment since almost all life on Earth depends on its action. This most abundant polypeptide in the world (Ellis, 1979), with a molecular composition of two subunits of 55 and 12.5 kDa (Kung, 1976), comprises 50% of leaf protein. Finally, this enzyme represents a protein, with nutri-tional potential, composed of amino acid building blocks deriving from the incorpora-tion of inorganic nitrogen (NH_4^+, NO_3^-) present in the soil. These important aspects motivated studies to validate the utilization of leaves as an alternative source of food protein (Barbeau, 1994; Lima *et al.*, 1998; Pedoni *et al.*, 1995). The best known pilot scale processes for production of leaf protein concentrate (LPC) were described for alfalfa leaves: (i) green LPC, ProXan I (Spencer *et al.*, 1971) and (ii) white LPC, ProXan II (Edwards *et al.*, 1975). In these processes, both fractions were separated by selective heat precipitation. The utilization of LPC in the human diet has remained almost non-existent while other newly developed protein concentrates and isolates from soy, cottonseed, peanut, whey, and other sources are already being incorporated into foods at a rapidly expanding rate (Friedman, 1996). Cassava (*Manihot esculenta* Crantz) is a starchy root that represents the staple food not only of the population from the north-east of Brazil but also from many countries in Africa. Its content of approximately 1% protein (Souci *et al.*, 1994) is not enough to supply human daily needs. Research on molecular biology has already been initiated in order to improve the biotechnological process for attaching protein, for instance from wheat to cassava (Clowes *et al.*, 1995). The development of a technological process for producing LPC of good quality is another approach for incorporating protein into cassava starchy food. Until now, the unique publication directed at separating and concentrating proteins from these leaves has focused on using ultrafiltration as the alternative technological process (Castellanos *et al.*, 1994). Moreover, the applicability of cassava leaves for fortification of biscuits has been done utilizing proteins obtained by a thermo-coagulation process (Meimbam *et al.*, 1982). Speculations based on nutritional aspects suggest the utilization of a plastein reaction for improvement of the protein digestibility of LPC (Rosas-Romero and Barata, 1987). However, the biggest limitation of utilizing the leaves of this crop is the presence of a potential health hazard component in their leaves and roots, the cyanogenic glycosides, which after enzymatic hydrolysis, catalysed by linamarase, produce hydrocyanic acid (Hughes *et al.*, 1992). Processes based on chemical dilution (washes) and enzymatic activity (heat treatment) are alternatives utilized for the liberation of HCN in root or leaves. As a consequence, these leaves are already utilized in Africa as a vegetable (Awoyinka *et al.*, 1995) and in Brazil in small communities in the form of flour, as described for rice leaves (Neelima *et al.*, 1978).

2. Technical approach for extracting proteins from cassava leaves

2.1 Characterization of the raw material production and varieties in Rio de Janeiro and Brazil

Although the annual production of cassava in Rio de Janeiro represents less than 1% (211 547 ton) of the total Brazilian production (24 539 638 ton) this crop is well appreciated and produced by different small planters in this city. Studies on the yield of LPC production and analyses of protein migration patterns on sodium dodecylsulphate–polyacrylamide gel electrophoresis (SDS–PAGE) were performed with crude protein fractions of leaves from cassava grown at two regions in Rio de Janeiro. In the winter, leaves of 10-month-old plants were produced from the experimental area of the agroecological field station at EMBRAPA-CNPAB and from a conventional producer, which supplies the best known and the most expensive cassava root for the central market of Rio de Janeiro. It is important to mention that the bank of cassava (*Manihot esculenta* Crantz) at EMBRAPA-CNPMF (Mato Grosso, Brazil) accounts for over 1641 accessions collected in several regions of Brazil and other countries. For this classification and division, parameters as botanical, agriculture and physico-chemical characteristics have been used (Fukuda *et al.*, 1997). No molecular markers have been used for the characterization of this crop in Brazil, as is already being done in others countries (Beeching *et al.*, 1993). Moreover, the majority of varieties utilized in Rio de Janeiro have not been catalogued yet and are designated by either the name of the region or the name of the planter that produces it. Additionally, this crop has a broad genetic diversity and the varieties are usually adapted to environmental conditions and consequently to several types of ambient under conditions of stress (Fukuda *et al.*, 1996).

2.2 Characteristics of environmental diversity (by soil analyses) of the two cassava producers in Rio de Janeiro

For an initial characterization of environmental diversity in the regions, the levels of the essential and non-essential minerals in the soil such as potassium, phosphorus and the aluminium were performed. Methods based on colorimetric analyses, titration and flame ionization were used (EMBRAPA, 1997). In *Figure 1* the distribution of phosphorus and aluminium in two depths of soils is shown.

Differences between the soils concerning organic matter levels and pH were observed. The agroecological and conventional farms utilized had, respectively, an organic matter of 12.2% and 1.2% and pH of 4.0 and 4.4. At this low pH, the hydrogen and the high levels of aluminium ions displace essential nutrients, interfere with the plant's biochemistry and may cause a toxic effect for vegetables (Hedin and Likens, 1996). The concentration of potassium (ppm) in the soil of the Conventional region was twice (76) that obtained for the agroecological region (33.5). These results already indicate the presence of altered abiotic factors that may influence adaptation of cassava to their environment conditions, specially concerned to the aluminium concentration in the soil of the 'conventional farm'. The concentration of aluminium in this soil was 15–33 times higher than that observed in the soil of the agroecological field in the Experimental Station of EMBRAPA. The amount of phosphorus in the soil of both producers differed only for soil from 40 cm of depth (*Figure 1*).

2.3 Development of the extraction method and mass balance for a bench process for obtaining leaf protein concentrate from cassava leaves

The isolation of protein from leaves, in general, has been mentioned as a process that shows more efficiency when performed at higher pH, which gives maximum yields at pH 12 (Betschart and Kinsella, 1973). However, the LPC obtained under those conditions are unattractive products that may be black and of low nutritive value. These brown pigments are called melanins which condense with proteins. These complex phenolic compounds are formed by non-enzymatic reaction of polymerization of *o*-quinones, which are produced by the reaction of polyphenol oxidase with mono- and

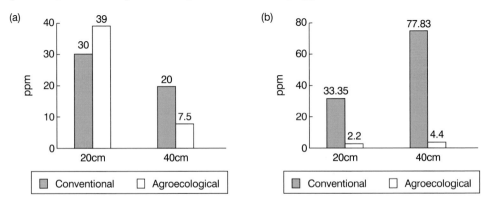

Figure 1. Distribution of (a) phosphorus and (b) aluminium into two depths of soil (20 and 40 cm) from two agriculture regions in Rio de Janeiro, conventional and agroecological.

Table 1. Yield of the bench process for isolating protein from leaves of cassava (*Manihot esculenta*), cultivated under conventional and agroecological conditions[a]

Fraction	Yield[a], wt%		
	Conventional[b]	Agroecological[c]	ProXan II (alfafa)[d]
White (LPC)	1.61	3.42	2.05
Green (LPC)	4.73	7.26	6.80
Fibrous residue	88.23	71.80	72.5
Liquid residue	5.13	11.97	15.45
Losses	0.30	0.30	3.2
Total	100	100	100

[a] Data on yield were obtained from freeze-dried products.
[b] One hundred g dried leaves corresponds to 366.7 g of fresh leaves.
[c] One hundred g dried leaves corresponds to 427 g of fresh leaves.
[d] ProXan II for alfafa.
LPC, leaf protein concentrate.

Table 2. Distribution of nitrogen obtained with the bench process for isolation protein from cassava (*Manihot esculenta*), cultivated under conventional and organic agriculture, in dry weight[a]

Item[b]	Total nitrogen content (g)		Distribution of total nitrogen (%)	
	Conventional	Agroecological	Conventional	Agroecological
Dried leaves	3.42	4.04		
Fibrous residue	2.68	1.92	78.36	47.52
Green fraction	0.26	0.94	7.60	23.27
White fraction[c]	0.07	0.17	2.05	4.21
Liquid residue	0.26	0.73	7.60	18.70

[a] Total nitrogen determination was performed by the combustion method using a LECO equipment model FP-428, as described by AOAC (1997).
[b] All fractions were freeze-dried and ground prior to analytical determination.
[c] Protein was quantified by the method of Bradford (1976) and nitrogen content was calculated by dividing the protein value by 6.25.

diphenols (Edwards *et al.*, 1975). *Figure 2* shows the bench process used for isolating protein from cassava. The extraction of total solids and nitrogen from leaves of cassava (agroecological) was, respectively, 22.65% and 45.54% (*Tables 1 and 2*). The results are show in *Tables 1* and 2. These values are twice higher than those observed for the leaves from the conventional agriculture. From the total nitrogen extracted (agroecological), 23.27% was determined to be in the green fraction, 4.21% in the white fraction and 18.70% in the residual liquid (*Table 2*). Interestingly, the amount of nitrogen present in the liquid residue represents almost 40% of the total nitrogen extracted. The absence of any precipitated material after treating this liquid with 5% and 10% trichloroacetic acid (TCA) indicates the absence of any proteic nitrogen. It is probable that the cyanogenic glycosides are contributing to this high nitrogen level. The yield of nitrogen extraction described in this paper for the cassava leaves from the Agroecological agriculture

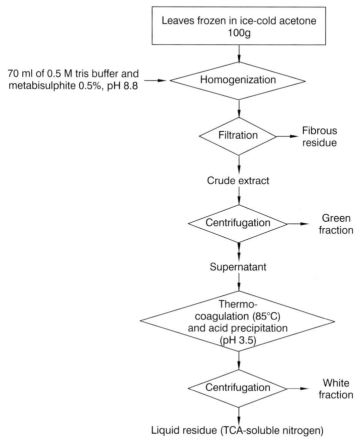

Figure 2. Bench process scheme for obtaining the green and white fraction of protein from cassava leaves

showed a similar percentage to that described for alfalfa leaves obtained by the ProXan II process (Edwards *et al.*, 1975).

2.4 Solubility and protein migration pattern of the two fractions (green and white) on SDS-PAGE

Both green and white fractions showed, after freezing and freeze-drying, total denaturation and no solubility in water and Tris-HCl buffer. Difficulties were also observed for making the fractions soluble after centrifugation and precipitation. Tentative to solubilizing these fractions in water, Tris-HCl 1 M pH 8.8 and NaCl 1 M were not successful. Better results were achieved with 6 M urea, which allowed the proteins to be kept in solution even after freezing. In *Figure 3* the electrophoreses of the white and the green fractions, obtained by SDS–PAGE according to the method of Laemmli (1974) revealed: (i) similar protein migration patterns for the green fraction extracted from leaves of the two regions (lanes 5 and 6) and (ii) differences between the protein pattern of the white fraction extracted from leaves of the two regions. Two bands of 32 kDa and

White fraction   Green fraction

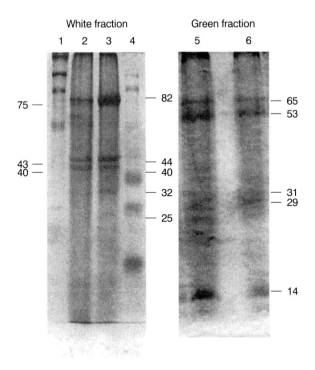

Figure 3. SDS–PAGE of the proteins of the white and green fractions from leaves of cassava, stained with coomassie brilliant blue R-250 (Sheen, 1991). Lane 1: molecular mass standards (208 kDa, myosin; 116 kDa, β-galactosidase; 84 kDa, bovine serum albumin; 47 kDa, ovalbumin); lane 4: molecular mass standards (104 kDa, phosphorylase B; 81 kDa, bovine serum albumin; 47.7 kDa, ovalbumin; 34.6 kDa, carbonic anhydrase; 28.3 kDa, soybean trypsin inhibitor; 19.2 kDa, lysozyme); lane 2: white fraction (conventional); lane 3: white fraction (agroecological); lane 5: green fraction (conventional); lane 6: white fraction (agroecological).

a slightly stained band of 25 kDa in the white fraction extracted from leaves of the Agroecological region were observed (lane 3).

Interestingly, these bands were not observed in the same fraction obtained from leaves of the Conventional agriculture. This difference may already represent an adaptation of this crop to the environment in this region (soil rich in aluminium). Moreover, the protein band of 25 kDa (lane 2) stains more intensively when ammoniacal silver staining is used (data not shown). The absence of similarity between the protein pattern of the white fractions extracted from leaves of the two different regions in Rio de Janeiro and the low yields of mass and nitrogen indicate that only the green fractions have the potential to be used as an additive for food fortification in industry. The possibility of development of a protein with allergenic potential may not be discarded. The production of proteins in response to biotic stress is a subject that has also been correlated to the allergenic effects in humans (Walter *et al.,* 1990). The protein pattern of the green fraction (lanes 5 and 6) resembles that expected for the enzyme RuBP (55 and 12.5 kDa), however with small differences of double banding at 65 and 53 kDa and 31 and 29 kDa. The protein pattern of the enzyme RuBP observed by Sheen (1991) in the white fraction of the different leaves was observed in the green fraction of cassava leaves in this work.

2.5 Speculations on the applicability of cassava leaves as an alternative source of protein in the food industry and as tool for identification of varieties adapted to environmental diversity

Initially, it is important to mention that the application of LPC from cassava leaves in the dried form is limited to the fortification of products because it is totally insoluble in water and, as a consequence loss of functional properties occurs. Nevertheless, the product can be stored as a cream under refrigeration, which permits technological properties, ideal for the preparation of soups like 'baby food' and sauces. Parallel studies carried out with spinach leaves have shown that the high levels of toxic nitrate present in these leaves are completely liberated to the residual liquid after isolation of the LPC. These results indicate that the polarity of the cyanogenic glycosides will also contribute to bring this compound to the residual liquid of cassava leaves processed for LPC production. The amino acid composition of the two LPCs is given in *Table 3*. These results show slightly lower contents of Ile, Leu, Thr and Val than those described for alfalfa LPC (Sheen, 1991). However, by comparing the essential amino acids of cassava LPC with the standard protein of FAO/WHO (1991), one concludes that all essential amino acids in cassava LPC are present in the above recommended amounts.

Table 3. Essential amino acid composition of leaf protein concentrate (LPC) (green fraction) from leaves of cassava cultivated in two different environmental systems[a] and from reference protein (FAO/WHO[b])

Amino acid (in g per 100 g protein)	Leaf protein concentrate (green fraction)		Reference protein FAO/WHO[b]
	Conventional	Agroecological	
Thr	3.23	3.37	2.85
Cys+Met	6.37	6.04	2.65
Val	4.44	4.39	3.20
Leu	7.66	7.90	5.55
Ile	4.10	4.15	2.88
Tyr+Phe	9.60	9.56	4.40
Lys	4.67	4.97	4.60
Trp	ND[c]	ND	1.05

[a] Samples were hydrolysed for 70 min (150°C) with 1 M HCl. Primary and secondary amine groups were derivatized respectively with o-phthaldialdehyde (OPA) and fluorenylmethyl chlorformate (FMOC). Analyses were performed in a HPLC system (Aminoquant) using reversed-phase column, hypersyl ODS 5 μm (20 cm × 2.1 cm). The following two buffers were utilized: (a) sodium acetate, 20 mM; triethylamine, 0.01 M and tetrahydrofuran; (b) methanol, 40%; acetonitrile, 40% and sodium acetate, 700 mM, pH 7.2.
[b] Data represent means of levels recommended by FAO/WHO (1991) for daily nutritional requirements of different ages.
[c] ND, not determined.

3. Conclusions

(i)   The green fraction is a potential product for fortification of food.
(ii) The white fraction is a potential tool for identification of environment diversity and
      for differentiation of the RuBP carboxylase form of cassava from others leaves.
(iii) The green fraction has good nutritional and technological characteristics.
(iv) Further biological assays must be conducted to certify the nutritional value of the
      proteins present in the green fraction.

Acknowledgements

The authors thank Dr Frederico Siqueira (EMBRAPA-CTAA) for analysis and interpretation of the aminogram, Dr Dejair de Almeida (EMBRAPA-CNPAB) for growth of
cassava and soil analysis, Dr Viktor C. Wilberg for allowing the utilization of the laboratory of pigments and the kindness of the producer of the best commercialized cassava in
Rio de Janeiro-Brazil, Sir Seizi Miyata. This work has been supported in part by grants
from the Brasilian National Council of Research.

References

Association of Analytical Chemists (1997) *Official Methods of Analysis*, 17th edn. AOAC,
    Washington, DC.
Awoyinka, A.F., Abegunde, V.O and Adewusi, S.R.A. (1995) Nutrient content of young cassava
    leaves and assessment of their acceptance as a green vegetable in Nigeria. *Plant Foods Human
    Nutr.* **47:** 21–28.
Barbeau, W.E. (1994) A historical prospective of leaf protein research and assessment of future
    needs. *Italian J. Food Sci.* **6:** 387–395.
**Beeching, J.R., Marmey P., Gavalda M.C., Noirot, M., Haysom, H.R., Hughes, M.A. and
    Charrier, A.** (1993) An assessment of genetic diversity within a collection of cassava (*Manihot
    esculenta*, Crantz) germplasm using molecular markers. *Ann. Bot.* **72:** 515–520.
Betschart, A. and Kinsella, J.E. (1973) Extractability and solubility of leaf protein. *J. Agr. Chem.* **21:**
    60–65.
Bradford, M.M. (1976) A rapid and sensitive method for quantitation of microgram quantities of pro-
    tein utilizing the principle of protein dye binding. *Analyt. Biochem.* **72:** 248–254.
Castellanos, R., Altamirano, S.B. and Moretti, R.H. (1994) Nutritional characteristics of cassava
    (*Manihot esculenta* Crantz) leaf protein concentrates obtained by ultrafiltration and acidic thermo-
    coagulation. *Plant Food Human Nutr.* **45:** 357–363.
Clowes, A.E.E., Tatham, A.S, Beeching, J.R. and Shewry, PR. (1995) Characterization of cassava
    root proteins. In: *The Cassava Biotechnology Network: Proceedings of the Second International
    Scientific Meeting* (eds A.-M. Thro and W. Roca). CIAT, Cali, Colombia, pp. 716–728.
Edwards, R.H., Miller, R.E., Fremery, D., Knuckles, B.E., Bickoff, E.M. and Kohler, G.O. (1975)
    Pilot, plant production of an edible white fraction leaf protein concentrate from alfalfa. *J. Agric.
    Food Chem.* **23:** 620–626.
Ellis, R.J. (1979) The most abundant protein in the world. *Trends Biochem. Sci.* **4:** 241–244.
EMBRAPA (1997) *Manual de métodos de análise de solo* (Centro Nacional de Pesquisa de Solos),
    2nd edn (eds M.E.C. Claressen and P.M. de Sinsa Magalhães).
FAO/WHO (1991) *Protein Quality Evaluations.* Food and Agriculture Organization of the United
    Nations. Rome, Italy, p. 66.

Friedman, M. (1996) Nutritional value of proteins from different food sources. A review. *J. Agric. Food Chem.* **44:** 6–29.

Fukuda, W.M.G., Silva, S. de O. and Porto, M.C.M. (1997) Caracterização e avaliação de germoplasma de mandioca (*Manihot esculenta* Crantz). Cruz das Almas, BA: EMBRAPA-CNPMF, 161 (catálogo).

Hedin, L.O. and Likens, G.E. (1996) Atmospheric dust and acid rain. *Sci. Am.* **275:** 56–60.

Hughes, M.A, Brown, K., Pancoro, A., Murray, B.S., Oxtoby, E. and Hughes, J. (1992) A molecular and biochemical analysis of the structure of the cyanogenic beta-glucosidase (linamarase) from cassava (*Manihot esculenta* Cranz). *Arch. Biochem. Biophys.* **295:** 273–279.

Kung, S.D. Tobacco fraction 1 protein: a unique genetic marker. *Science* **191:** 429–434.

Laemmli, U.K. (1970) Cleavage of structural proteins during the assembly of the head of bacteriophage t4. *Nature* **227:** 680–685.

Lima, F.H., Wilberg, V.C., Matsuura, F., Almeida, D.L. and Stephan, M.P. (1998) Avaliação comparativa do potencial de utilização de folhas de hortaliças tuberosas para obtenção de isolados proteicos. *Anais do XVI Congresso Brasileiro de Ciência e Tecnologia de Alimentos* **2:** 1073–1076.

Meimbam, E.J., Bautista, J.G. and Soriano, M.R. (1982) Studies on the fortification of biscuits with cassava leaf protein concentrates. *Philipp. J. Nutr.* **35:** 82–86.

Neelima, N., Meera, K. and Punekar, B.D. (1978) Improvement of the nutritive quality of rice through fortification of dehydrated leafy powders. *Indian J. Nutr. Dietetics* **15:** 346–349.

Pedoni, S., Selvaggini, R. and Fantozzi, P. (1995) Leaf protein availability in food: significance of the binding of phenolic compounds to ribulose-1,5-diphosphate carboxylase. *Lebensmittel Wissenschaft Technologie* **28:** 625–634.

Rosas-Romero, A.J. and Barata, C. (1987) Composition, functional properties and biological evaluation of a plastein from cassava leaf protein. *Plant Foods Human Nutr.* **37:** 85–96.

Sheen, S.J. (1991) Comparison of chemical and functional properties of soluble leaf proteins from four species. *J. Agric. Food Chem.* **39:** 681–685.

Souci, S.W., Fachmann, W. and Kraut, H. (1994) *Food Composition and Nutrition Tables*, 5th revised completed edn. Medpharm GmbH Scientific Publishers, Stuttgart, Germany.

Spencer, R.R., Mottola, A.C., Bickoff, E.M., Clark, J.P. and Kohler, G.O. (1971) The PRO-XAN process: the design and evaluation of a pilot plant system for the coagulation and separation of the leaf protein from alfalfa juice. *J. Agric. Food Chem.* **19:** 504–507.

Walter, M.H., Liu, J.W., Grad, C., Lamb, C.J. and Hess, D. (1990) Bean pathogenesis-related (PR) proteins deduced from elicitor-induced transcripts are members of a ubiquitous new class of conserved PR proteins including pollen allergens. *Mol. Gen. Genet.* **222:** 353–360.

Index